NORTH CAROL
STATE BOARD OF COMMU
LIBRARIES
SAMPSON COMMUNITY COLLEGE

A SYSTEM FOR ADVANCEMENT IN THE FIRE SERVICE

BY

ARTHUR R. COUVILLON

INFORMATION GUIDES, HERMOSA BEACH, CALIFORNIA

A SYSTEM FOR ADVANCEMENT IN THE FIRE SERVICE

1990

BY

ARTHUR R. COUVILLON

FIRST EDITION
ALL RIGHTS RESERVED.
THIS BOOK, OR PARTS THEREOF,
MAY NOT BE REPRODUCED IN ANY FORM
WITHOUT PERMISSION OF THE PUBLISHER
Printed in the United States of America
COPYRIGHT, 1990 BY INFORMATION GUIDES

Library of Congress Cataloging Data:

Couvillon, Arthur R.
   A System For Advancement In The Fire Service
1. Fire Service---Handbooks, manuals, etc.
2. Fire Engineer--Handbooks, manuals, etc.
3. Fire Captain---Handbooks, manuals, etc.
4. Information----Handbooks, manuals, etc.
5. Promotional----Handbooks, manuals, etc.
I. Title
LCCN 88-83385
ISBN 0-938329-56-1

INTRODUCTION

The intent of this book is to provide a list of information pertaining to the duties, knowledge, responsibilities, training, and education that is necessary in order to successfully promote in the Fire Service

The goal of this book is to show promotional candidates a program that they may use to successfully promote in the Fire Service.

A vast number of Cities across the U.S.A. give promotional exams for the positions of **FIRE ENGINEER, APPARATUS DRIVER, PUMP OPERATOR, FIRE CAPTAIN, FIRE LIEUTENANT, ETC.** Most Fire Departments encourage all of their personnel to improve their abilities and thus become more valuable assets to the departments and to the communities which they protect.

ACKNOWLEDGEMENTS

Appreciation is expressed to those individuals, fellow Firefighters, Fire Engineers, Fire Captains, Fire Training Officers, & Fire Science Instructors that have contributed to the compilation of this "System For Advancement In The Fire Service"

ABOUT THE AUTHOR

The author is a veteran Firefighter, with 18 years of Firefighting experience. Art shares the knowledge that he has gained during this period with a series of published Fire Service Study guides:

A SYSTEM FOR A CAREER IN THE FIRE SERVICE

FIREFIGHTER WRITTEN EXAM STUDY GUIDE

FIREFIGHTER ORAL EXAM STUDY GUIDE

A SYSTEM FOR ADVANCEMENT IN THE FIRE SERVICE

FIRE ENGINEER WRITTEN EXAM STUDY GUIDE

FIRE ENGINEER ORAL EXAM STUDY GUIDE

FIRE CAPTAIN WRITTEN EXAM STUDY GUIDE

FIRE CAPTAIN ORAL EXAM STUDY GUIDE

All books are available from: Information Guides, P. O. Box 531, Hermosa Beach, CA 90254

TABLE OF CONTENTS

PAGES

SECTION 1 :   **1-5**

PRIOR TO OBTAINING A POSITION AS A FIREFIGHTER
- ENTRANCE FIREFIGHTER   2-4
- REQUIREMENTS   2
- THINGS TO ESTABLISH   2
- THINGS TO LEARN   2
- EXPERIENCE   3
- EDUCATION   3
- EXAM CHECK LIST   4
- FIREFIGHTER ENTRANCE EXAM QUESTIONS   5

SECTION 2 :   **7-17**

AFTER OBTAINING A POSITION AS A FIREFIGHTER
- RESPONSIBILITIES   8
- JOB PERFORMANCE   8
- QUALIFICATIONS   9
- JOB KNOWLEDGE   10
- EXPERIENCE   10
- FIRE ENGINEER EXAM QUESTIONS   11-17

SECTION 3 :   **13-48**

AFTER PROMOTING TO THE POSITION OF FIRE ENGINEER
- RESPONSIBILITIES   20
- JOB PERFORMANCE   20
- JOB KNOWLEDGE   20
- EXPERIENCE   20
- FIRE CAPTAIN EXAM QUESTIONS   21-30
- ASSESSMENT CENTERS   31-32
- SIMULATOR EXAMS   33-48

SECTION 4 :   **49-50**

AFTER PROMOTING TO THE POSITION OF FIRE CAPTAIN
- RESPONSIBILITIES   50
- JOB PERFORMANCE   50
- JOB KNOWLEDGE   50
- EXPERIENCE   50

SECTION 5 :   **51-78**

EXAM ELEMENTS AND STRATEGY
- EXAM PROCTORS   52
- EXAM FORMATS   53-60
- CANDIDATES   61
- PREPARATION   62-78

SECTION 6 :   **79-148**

FIRE AND FIREFIGHTING PRINCIPLES
- FIRE SCIENCE   80-93
- FIREFIGHTING   94-114
- WATER SUPPLY   114-123
- EXTINGUISHING SYSTEMS   124-135
- FIRE STREAMS   135-148

SECTION #1

PRIOR TO OBTAINING A POSITION AS A FIREFIGHTER

ENTRANCE FIREFIGHTER

Advancement in the Fire Service actually begins with your preparation for a career in the Fire Service and continues even when you have reached your own personal highest expectations.

During the period that you are preparing for a career in the Fire Service you should:

Be aware of entrance **REQUIREMENTS:**

1. High School graduate.
2. 18 years of age.
3. Meet medical and physical requirements.
4. Information on job announcement.

Initiate and **ESTABLISH** the following:

1. Good study habits.
2. Good work habits.
3. Your own personal Fire Service Library.

Things to **LEARN:**

1. Job requirements.
2. Information on job flyer.
3. How to file for exams.
4. How to relate in Fire Service situations.
5. How to relate with others.
6. How to apply knowledge that you learn.
7. How to be flexible.
8. How to be a "team player".
9. How to learn.
10. To read instructions.
11. To follow directions.
12. Fire Service chain of command.

13. Your personal weaknesses.
14. Your personal strong points.
15. Resume writing.
16. Duties and responsibilities of a Firefighter.
17. Rescue techniques.
18. Firefighting techniques.
19. Fire Prevention techniques.
20. Manual skills.
21. Foreign language, such as spanish, ETC.
22. As much about the Fire Service as you can.

Acquire **EXPERIENCE** in:
1. Fire Service written exams.
2. Fire Service oral exams.
3. Fire Service "Cadet" program or Volunteer F.S.
4. Fire station atmosphere. (visit fire stations)

EDUCATION, enroll in:
1. Fire Science courses:
2. Fire Service orientation.
3. Fire apparatus and procedures.
4. Fire hydraulics.
5. Building construction.
6. First aid and CPR.
7. Rescue procedures.
8. E.M.T. 1
9. State Fire Service courses:
   Fire Administration, Management/Supervision.
   Fire Prevention, Arson Investigation.
   Fire Command, Instructor courses. ETC
10. Fire prevention techniques.
11. Fire Service periodical subscriptions.
12. Fire Service Academy.

Establish an **EXAM CHECK LIST:**
1. Communicate with Personnel Department.
2. Communicate with Fire Department.
3. Get advice from Fire Science instructors.
4. Investigate for job openings in newspaper.
5. Acquire job specifications.
6. Acquire the exam announcement.
7. Obtain the jobs medical standards.
8. Get the job application.
9. Complete the job application.
10. Formulate a resume.
11. Submit the job application.
13. Submit the resume.
14. Prepare for written exam.
15. Prepare for physical agility exam.
16. Prepare for oral exam.
17. Learn location of all portions of exam.
18. Make visits to the Fire Stations.
19. Research particulars of the Fire Department.
20. Research the particulars of the community.
21. Take practice oral exams.
22. Select clothing for the oral interview.

Read fire Service related books and magazines.
Acquire a Fire Science college degree.
Get involved in civil service activities.
Stay healthy:
1. Proper diet.
2. Exercise.

For a complet guide on how to become a Firefighter, read: **"THE COMPLETE FIREFIGHTER CANDIDATE"**, available from: INFORMATION GUIDES.

FIREFIGHTER EXAM QUESTIONS:

Some examples of entrance level Fire Service written exam subjects that you should be familiar with include:

1. Mechanical knowledge.
2. Math and the concepts of math.
3. Number sequences.
4. Matching figures, forms etc.
5. Pattern evaluation.
6. Reading comprehension.
7. Reading judgement.
8. General knowledge.

For a complete guide to Firefighter written exams, read: **"FIREFIGHTER WRITTEN EXAM STUDY GUIDE"**, available from: INFORMATION GUIDES.

Some examples of entrance level oral exam questions that you should be familiar with are:

1. Tell us something about yourself.
2. How have you prepared yourself for the position of Firefighter?
3. Why do you want the position of Firefighter?
4. What are your qualifications for the position of Firefighter?
5. Why should you be the first candidate chosen?
6. What are your immediate and ultimate goals in the Fire Service?
7. When did you decide to become a Firefighter?
8. What is your most outstanding asset?
9. What is your biggest weakness?
10. Do you have any job application for other Fire Departments?
11. Is there anything that you would like to add?
12. ETC.

For a complete guide to Firefighter oral exams, read: **"FIREFIGHTER ORAL EXAM STUDY GUIDE"**, available from: INFORMATION GUIDES.

SECTION #2

AFTER OBTAINING A POSITION AS A FIREFIGHTER

AS FIREFIGHTER

Be aware and perform the duties and RESPONSIBILITIES of a Firefighter:

1. Maintenance of:
    Station.
    Fire hydrants.
    Alarm boxes.
    Apparatus and equipment.

2. Respond to emergency incidents.

3. Perform emergency procedures.

4. Perform rescue techniques.

5. Public relations:
    Tours.
    Demonstrations.

6. Take part in:
    Training.
    Fire prevention.
    Pre-fire planing.

7. Be available and ready to act as Fire Engineer.

8. Continue studying.

9. Related duties as needed.

Firefighter will be able to identify:

1. Organization of the Fire Department.

2. Size and scope of the Fire Departments operations.

3. Standard Operating Procedures of the Fire Department.

4. Rules and regulations.

Display confidence in **JOB PERFORMANCE** including:

1. Fire Station maintenance.

2. Apparatus and Equipment maintenance.

3. Firefighting.

4. Company fire inspections/Pre-fire planning.

5. Related duties as required.

Firefighters should be aware of NFPA 1001 Firefighter professional **QUALIFICATIONS:**

Set aside time for self study every shift that you are on duty.

Do the best job that you are capable of doing.

Make it obvious to your peers and supervisors that you are pleased to be a Firefighter.

Make yourself available for special details, assignments and projects including:

1. Hydrant maintenance.
2. Pump and Hose testing.
3. Training and instruction.
4. Tours and Demonstrations.
5. Equipment fabrication.

Come up with new ideas for improving your Fire Department.

Bid for as many varied assignments that your Fire Department makes available:

1. Engine pumper companies.
2. Truck companies.
3. Salvage companies.
4. Rescue companies.
5. Paramedics.
6. Dispatch.
7. Fire prevention.
8. High incident areas.
9. Harbor districts.
10. High rise areas. ETC.

Gain **KNOWLEDGE** in:

1. Fire apparatus, tools and equipment.
2. Fire pumps and streams.
3. Fire prevention.
4. Fire behavior.
5. Fire hydraulics.
6. Water supply.
7. Hazardous materials.
8. Fire extinguishing systems.
9. Automatic and mutual aid. (response assistance)

Gain **EXPERIENCE** in:

1. Fire Department responses.
2. Driving apparatus.
3. Pumping fire pumpers.
4. Report writing.
5. Performing at emergency incidents:
6. Hazardous incidents.
   Fires.
   Rescues.
   Entrapment.
   Hazardous materials.

Learn the duties of your Fire Departments next two ranks up from your present job classification:

Fire Engineer, Apparatus Driver, Pump Operator.

Fire Captain, Company Officer, Lieutenant.

Participate in your Fire Departments certification program that allows you to "act" in those positions.

Continue to establish, add to and complete the list that was started while preparing for a career in the Fire Service.

Be aware of the testing procedures for the next promotional exam.

Study specifics for the next promotional exam.

ENGINEER EXAM QUESTIONS:

Some examples of typical written questions for the Fire Engineers exam:

EXAMPLE #1

QUESTION:

What is the basic purpose of a pressure radiator cap?

CHOICES:

A. To force steam pressure out of the blow-off valve.

B. To raise the boiling point of the water within the radiator.

C. To prevent a vacuum within the cooling system.

D. To force the water through the cooling system under pressure.

ANSWER = "B"

EXAMPLE #2

QUESTION:

While drafting with a centrifugal pump, what will be the result of a clogged strainer ?

CHOICES:

A. Pump noise.

B. Pump engine stalling.

C. Pump vacuum gauge will register towards zero.

D. Pump vacuum gauge will show an increase in vacuum.

E. Head velocity will increase.

ANSWER = "D"

EXAMPLE #3:

QUESTION:

Radiation of heat means ?

CHOICES:

A. Movement of heat by air ducts.

B. Transmission of heat by air currents.

C. Transmission of heat in all directions, from the center.

D. Transmission of heat by direct contact.

ANSWER = "C"

EXAMPLE #4:

Written exams will include questions about:
1. Apparatus and equipment.
2. Fire pumps and streams.
3. Water supply and extinguishing systems.
4. Fire hydraulics.
5. Fire behavior and firefighting.
6. Fire prevention.
7. Hazardous materials.
8. Pumping techniques.
9. Mechanics, general information, ETC.

For a complete guide to Fire Engineer written exams read: **"FIRE ENGINEER WRITTEN EXAM STUDY GUIDE"**, available from: INFORMATION GUIDES.

Some examples of typical practical Fire Engineer exam questions, with possible responses:

EXAMPLE #1

QUESTION:

Explain how the **AUXILIARY COOLER** (heat exchanger) on your Fire Departments apparatus works. The apparatus that you are assigned to.

POSSIBLE RESPONSE:

Water comes from the discharge side of the apparatus pump and travels through a series of plumbing (copper tubing) to a heat exchanger unit. The heat exchanger unit has a series of copper tubes within its casing. The cool water from the pump travels through these tubes. Also within the casing of the heat exchanger there is water circulating from the engines cooling system. The water from the pump and the water from the engines cooling system never intermix. The water from the pump cools the water from the engines cooling system by convection. The water from the tubes within the heat exchanger continues back to the suction side of the pump.

EXAMPLE #2

QUESTION:

What device is used to bypass water between the suction and discharge allowing a positive displacement pump to operate while the discharge gates are closed?

POSSIBLE RESPONSE:

The device used in a positive displacement pump that allows operation while the discharge gates are closed is a **CHURN VALVE**.

EXAMPLE #3

QUESTION:

What are some of the causes of low pump capacity?

POSSIBLE RESPONSE:

Causes of low pump capacity :
 1. Low engine horsepower.
 2. Engine overheated.
 3. High altitude.
 4. Improperly set pressure device.
 5. Wrong pumping gear.
 6. Air leaks.
 7. Restricted suction.
 8. Aerated water.
 9. High water temperature.
10. Leaks on the suction side of the pump.
11. Limited water supply.
13. Foreign matter in impellers.
14. Worn clearance rings.

EXAMPLE #4

Some of the things to be prepared for during the practical are:

1. Apparatus driving:
    Apparatus specifications.
    Equipment checks.
    Engine starting.
    Take off and gear selection.
    Steering and turns.
    Road operation.
    Application of road laws.
    Braking and stopping.
    Backing.
    Securing apparatus.

2. Pumping operations:
    General knowledge of Depts. pumping apparatus.
    Knowledge and use of pumps and controls.
    Knowledge of equipment on Depts. apparatus.
    Pump operations at a hydrant and at draft.
    Accurate calculations (field hydraulics)
    Apparatus maintenance.

3. Truck operations:
    Knowledge of Departments truck apparatus.
    Truck specifications.
    Location of equipment and their use.
    Types and sizes of ladders.
    Knowledge of other special tools.
    Truck operations.
    Turn-table operation.
    Water tower operation.
    Stowing of ladder.

Some examples of typical oral interview questions, with possible responses, for the Fire Engineer exam:

EXAMPLE #1

QUESTION:

In your opinion, what is the importance of the Fire Engineers position ?

POSSIBLE RESPONSE:

Some of the things that make the position of Fire Engineer an important position are:

1. The safe driving and operation of apparatus and equipment.

2. The proper maintenance of apparatus and equipment.

3. The added responsibilities of the knowledge and experience required for the position.

15

EXAMPLE #2

QUESTION:

How have you prepared yourself for the position of Fire Engineer?

POSSIBLE RESPONSE:

I have prepared myself for the position of Fire Engineer through motivation and involvement in the Fire Department. I have also prepared myself through training and motivation along with reading and serious study. During my "4" years of experience as a Firefighter and "2" years as a Paramedic, I have gained the necessary experience and knowledge for the position of Fire Engineer.

BE PREPARED TO GIVE EXAMPLES AS TO HOW YOU ARE PREPARED TO HANDLE THE RESPONSIBILITIES OF THE POSITION!

                              EXAMPLE #3

QUESTION:

Why do you want the position of Fire Engineer

POSSIBLE RESPONSE:

I want the position of Fire Engineer for the new and different challenges, along with the increased responsibilities of driving, operating, and proper maintenance of apparatus and equipment. The position of Fire Engineer will offer me the opportunity to improve myself and the department, by allowing me to apply my knowledge and experience.

EXAMPLE #4

QUESTION:

What are the responsibilities of a Fire Engineer spotted at a hydrant, pumping at a fire ?

POSSIBLE RESPONSE:

The responsibilities of a Fire Engineer spotted at a hydrant, pumping at a fire, are:

1. Safety of personnel.
2. Proper use of apparatus and equipment.
3. To be aware of how engine is performing:
4. Engine temperature.
5. Oil pressure.
6. Engine RPM in relation to discharge output.
7. Engine pressure on all discharge gates.
8. Amount of water being used and available.
9. Other pumpers and apparatus.
10. Hydrant locations, water sources.
11. Other apparatus on scene.
12. Other apparatus responding to the scene.
13. Equipment available and being used.
14. Radio communications.
15. Location of hose lines and nozzle pressures:
    Inside.
    Outside.
    Basement. (lower levels)
    Ground floors.
    Upper elevations.

For a complete guide to the Fire Engineer oral exam, read: **"FIRE ENGINEER ORAL EXAM STUDY GUIDE"**, available from: INFORMATION GUIDES.

SECTION #3

AFTER PROMOTING TO THE POSITION OF FIRE ENGINEER

AS FIRE ENGINEER

Be aware of and perform the **RESPONSIBILITIES** and duties of the position of Fire Engineer:

1. Be aware of Fire Department policy & procedures
2. Be aware of Fire Department rules/regulations.
3. Safe & proper driving of apparatus & equipment.
4. Safe & proper operation of apparatus and equipment.
5. The use of good judgement & remain alert.
6. Proper maintenance of apparatus & equipment.
7. Related work as required & assigned.
8. Through knowledge of city geography.
9. To stay current of:
    Water supply and hydrants.
    Firefighting techniques.
    First aid techniques.
10. To be involved in:
    Fire prevention.
    Pre-fire plans.
    Hydrant maintenance.
    ETC.

Continue to follow and complete the two previous sections as far as **JOB PERFORMANCE**, intensify your **TRAINING, EXPERIENCE,** and **KNOWLEDGE** in areas such as:

1. Management.
2. Supervision.
3. Leadership.
4. Planning.
5. Public relations.
6. Report writing, records keeping.
7. Staff assignments
8. Budgeting.
9. Fire investigation.
10. New procedures.

Add training in assessment centers and simulator exercises.

CAPTAIN EXAM QUESTIONS:

Some examples of typical written questions for the Fire Captain exam:

EXAMPLE #1

QUESTION:

In order to gain the greatest results, organization should be considered ?

CHOICES:

A. Primary objective.

B. Means to an end.

C. An important established order.

D. Time conserving mechanism.

ANSWER = "B"

EXAMPLE #2:

QUESTION:

What is the primary goal of a budget review ?

CHOICES:

A. To establish the necessary fiscal controls.

B. To prescribe the minimum expenditures.

C. To audit prior expenditures.

D. To determine the minimum funds available for operations at a given level.

ANSWER = "D"

EXAMPLE #3

QUESTION:

In investigating a suspicious fire, the first thing to ascertain is the ?

CHOICES:

A. Ignition source.

B. Point of origin.

C. Route of fire extension.

D. Possibility of arson.

ANSWER = "B"

EXAMPLE #4

Fire Captain written exam will cover such areas as:

1. Fire administration.
2. Firefighting and Fire behavior
3. General fire knowledge.
4. Hazardous materials.
5. Fire prevention, Arson.
6. Training.
7. Supervision, Management, Leadership.
8. Planning.
9. Morale.
10. Delegation.
11. Rules and regulations, Policies and procedures.
12. Public relations.
13. Rescue techniques.
14. Extinguishing systems.
15. Reports and records.
16. Building construction.
17. Apparatus and equipment. ETC.

For a complete guide to the Fire Captain written exam read: **"FIRE CAPTAIN WRITTEN EXAM STUDY GUIDE"**, available from: INFORMATION GUIDES.

Some examples of typical oral interview questions, with possible responses, for the Fire Captain exam.

EXAMPLE #1

QUESTION:

Tell us about YOURSELF.

POSSIBLE RESPONSES:

I am "28" years old, married/unmarried with "2" children. I have been a Firefighter with the "City" Fire Department for "8" years of which "3" years I was a Paramedic, and "2" years as a Fire Inspector, currently I am a Fire Engineer. During these years, through MOTIVATION and INVOLVEMENT in the Department, I have gained EXPERIENCE and the KNOWLEDGE necessary to prepare myself for the position of Fire Captain and/or Fire Lieutenant. SUCH AS:

Eight years as a Firefighter:

1. Firefighting techniques.//2. Company fire prevention techniques.//3. Public relations.//4. Rules and regulations.//5. Station maintenance.//6. Reports and records.//7. Hydrant maintenance.

Three years as a Paramedic:

1. Normal and emergency driving situations.
2. Responsibility of emergency situations.
3. Decision making.
4. Responsibility of apparatus and equipment.
5. Rescue techniques.
6. Record keeping and reports. (paper work)
7. Delegation of task.

Two years in Fire Prevention Bureau:
1. Responsibility of fire prevention areas.
2. Familiarization of codes and ordinances.
3. Public relations.
4. Planning.
5. Reports and records.
6. Arson.
7. Training.

Two years as Fire Engineer:
1. Apparatus driving/equipment maintenance.
2. Responsibility of apparatus and equipment.
3. Familiarization of apparatus and equipment.
4. Familiarization of water supply.

Currently in the position of Fire Engineer.
Currently on the Fire Captains Promotional List.

EXAMPLE #2

QUESTION:

How have you prepared yourself for the position of Fire Captain and/or Fire Lieutenant ?

Describe the motivation, education, training and experience you have had to qualify you to perform the duties of a Fire Captain and/or Fire Lieutenant.

POSSIBLE RESPONSE:

I have prepared myself for the position of Fire Captain through motivation and involvement in the "City" Fire Department. I have also prepared myself through training and education along with reading and serious study. During my "8" years as a Firefighter, "3" years as a Paramedic, "2"years as a Fire Engineer, and "2" years as a Fire Inspector, I have gained the necessary experience and knowledge for the position of Fire
Captain and/or Fire Lieutenant.
SEE PREVIOUS RESPONSE!
GIVE EXAMPLES OF PARTICULARS THAT APPLY TO YOU!

EXAMPLE #3

QUESTION:

How have you contributed to your Fire Department?

POSSIBLE RESPONSE:

During the "8" years that I have been a Firefighter with the "City" Fire Department I have contributed in many ways:

1. Involved in designing and fabricating ways to carry various tools and equipment on various apparatus.

2. Involved in designing and writing specifications for fire and rescue equipment and apparatus.

3. Researched and budgeted for various items of tools, equipment and apparatus.

4. Prepared lesson plans for various tools, equipment, apparatus, fire prevention techniques, etc.

5. Involved in safety projects.

6. Involved in various maintenance and overhaul of apparatus and equipment.

7. Prepared training manuals for the "City" Fire Departments Fire Engineer and Fire Captain certification programs.

8. Prepared training manual for the proper operating procedures of the departments ladder trucks.

9. Prepared training manual for the proper procedures of departments Fire Prevention program.

10. Projected a positive image of the department while giving lectures and demonstrations for various equipment and apparatus to various groups within the city.

11. Participated in the departments Paramedic program as a Paramedic and as an instructor for CPR and the departments EMT training program.

12. Participated in the departments school training program.

ADD/DELETE ANYTHING THAT MAY/MAY NOT APPLY FOR YOU!
BE PREPARED TO GIVE SPECIFICS FOR ANY OF THE ABOVE!

EXAMPLE #4

QUESTION:

Why do you want the position of Fire Captain and/or Fire Lieutenant?

POSSIBLE RESPONSE:

I want the position of Fire Captain and/or Fire Lieutenant for the new and different challenges, along with the increased responsibilities that the position assumes in relation to firefighting, fire prevention, supervision, and other task. The position will offer me the opportunity to improve myself and the department by allowing me to apply my knowledge and experience.

ADD ANYTHING THAT MAY BE APPLICABLE FOR YOU!

EXAMPLE #5

QUESTION:

What are your qualifications for the position?

POSSIBLE RESPONSE:

My qualifications extend from five distinct sources:

1. Motivation
2. Training
3. Education
4. Experience
5. Job knowledge

MOTIVATION:

I have a strong desire to improve myself and the "City" Fire Department through involvement in various department projects.

GIVE EXAMPLES!

TRAINING:

My training includes reading and very serious study along with inside and outside department drills and classes, including State Fire Officer Training.
GIVE EXAMPLES!

EDUCATION:

My education includes a College degree in Fire Science with 28 unites of Fire Science. Currently I am enrolled in various Fire Science courses.

GIVE EXAMPLES!

EXPERIENCE:

Give the number of years that YOU have served in the Fire Service, along with the type of experience YOU have gained during this period of time such as:

1. Firefighter
2. Fire Paramedic
3. Fire Inspector
4. Arson Investigator
5. Fire Engineer
6. Acting Fire Captain

HAVE BEEN RESPONSIBLE FOR:

1. Apparatus and equipment.
2. Subordinates.
3. Emergency situations.
4. Small and large fires.
5. Haz Mat incidents.
6. All phases of Fire Prevention.
7. Training of subordinates.
8. Quarterly task.
9. Delegation of task.
10. Reports and records.
11. Assignment of personnel and task.
12. Morale.
13. Management and supervision.
14. Demonstrations.
15. Rules and regulations.
16. Arson investigation.

ADD OR DELETE ANY EXAMPLES THAT ARE/NOT APPROPRIATE!

JOB KNOWLEDGE:

I have a through knowledge of the requirements for the position:

1. Management and organization.
2. Supervision and leadership.
3. Orders and delegation.
4. Planning.
5. Morale.
6. Reports and records.
7. Training.
8. Firefighting.
9. Fire prevention.
10. Target hazards.
11. Water supply.
12. City geography.
13. Department policy and procedures.
14. Department rules and regulations.
15. ETC.

ADD OR DELETE ANY EXAMPLES THAT APPLY TO YOU!

EXAMPLE #6

QUESTION:

What kind of work environment are you the most comfortable?

POSSIBLE RESPONSE:

I am comfortable in many different kinds of environments. I am the most comfortable in an environment where I have the opportunity to work and interact with people.

ADD/DELETE WHATEVER IS APPROPRIATE FOR YOU

EXAMPLE #8

QUESTION:

As Fire Captain, what is the first thing that you would teach a new recruit?

POSSIBLE RESPONSE:

As Fire Captain, the first thing that I would teach a new recruit is the concept of team work, no room for free lance operations in the Fire Service, we must operate as a team to accomplish the task given, provide for maximum safety for all and to be successful in all areas of a Firefighters responsibilities.

EXAMPLE #9

QUESTION:

What are the four major forms of "strategy" and how are they utilized?

POSSIBLE RESPONSE:

The four major forms of "strategy" are:

1. OFFENSIVE: utilized when the fire is small or when an attack is made directly on the seat of the fire. (aggressive, close attack, followed-up with other forces for support)

2. DEFENSIVE: utilized by protecting exposures without the advancing of heavy hose streams.

3. OFFENSIVE-DEFENSIVE: stop extension of fire with back-up support after an aggressive fire attack first on the major involvement of fire. ("blitz attack" = making a strong fire attack while setting-up defensive lines)

4. DEFENSIVE-OFFENSIVE: start by taking holding actions to protect exposures, then utilize incoming units, equipment, and manpower to initiate an aggressive fire attack.

EXAMPLE #10

QUESTION:

What are the eight basic divisions of "fire strategy?

POSSIBLE RESPONSE:

The eight divisions of basic "fire strategy" are:
1. Size-up.
2. Rescue.
3. Extinguishment.
4. Ventilation.
5. Salvage.
6. Confinement.
7. Exposure.
8. Overhaul.

EXAMPLE #11

QUESTION:

As Fire Captain, you observe a Firefighter performing a task improperly. When you discuss this with him, you are told that his previous Fire Captain taught him this method. What would be the best action to take at this point?

POSSIBLE RESPONSE:

In this situation as Fire Captain it would be best for me to indicate to the Firefighter as to why the method that he was using is not acceptable.

For a complete guide to the Fire Captain oral exam, read: **"FIRE CAPTAIN ORAL EXAM STUDY GUIDE"**, available from: INFORMATION GUIDES.

FIRE SERVICE ASSESSMENT CENTERS

Fire Service Assessment Centers are a thorough and extensive technique for enhancing a Fire Departments precision in measuring existing and/or possible aptitude of a candidate for promotion.

Fire Service Assessment Centers are directed in an authentic type of presentation so that the skills for the position will manifest to the proctors by the candidates behavior.

Fire Service Assessment Centers normally take one full eight hour day to complete. (sometimes two or three eight hour days)

Assessment of promotional candidates are usually preceded by a training session for the assessors.

Assessors are selected by your particular Fire Department and the personnel office.

Assessors will observe you in the exercise and systematically rate your performance.

Assessors are normally instructed to base their ratings and reports only on what takes place within the assessment center.

During the Assessment Center don't try to be someone other than yourself! Relax and get into the exercise by putting yourself into each situation and reacting as you normally would, not as you conceive the assessors would want you to.

Assessors will be watching your performance in terms of many various skills:

1. Oral/written communication.

2. Problem analysis.

3. Planning and organization.

4. Independence.

5. Interpersonal relations.

6. Organizational sensitivity.

7. Development of subordinates.

8. Persuasiveness.

9. Judgement.

10. Oral presentation.

The basic requirements for any management position will include skills in:

1. Planning and organizing.
2. Coordinating.
3. Leadership.
4. Budgeting.
5. Public relations.
6. Employee relations.
7. Personnel and personal development.
8. Management.

For a complete guide to Fire Service Assessment Centers, read: **"FIRE CAPTAIN ORAL EXAM STUDY GUIDE"**, available from : INFORMATION GUIDES.

INCIDENT SIMULATOR EXAMS

BASIC INFORMATION CONCERNING SIMULATOR EXAMS

WHAT IS A SIMULATOR EXAM?

SIMULATOR EXAMS are a technique of accessing a candidates fire command practices and principles within a classroom simulator environment with the use of mechanical and audio-visual equipment producing an emergency incident. The technique evaluates the candidates size-up and fire ground tactics and strategy within a controlled environment. The test allows for various fire situations to be created and modified in response to the actions taken by the candidate along with role players. (role players explained later)

WHY HAVE SIMULATOR EXAMS?

A simulator can create various situations for the candidate that could actually take place during a real incident.

Fire situations can be modified and predicted in response to the actions taken by the candidate and role players

The simulator exam can evaluate the candidates:

1. Command skills.
2. Control and management skills.
3. Knowledge of structural conditions.
4. Knowledge of fire behavior factors as they relate to the spread and growth of fire.
5. Knowledge of target hazards.
6. Fire ground procedures.
7. Decision making methods.
8. Communication skills.
9. Radio procedures.

SIMULATOR TECHNIQUES:

Simulators incorporate the use of overhead projectors and 35 mm projectors for the reproduction of fire and or smoke. Also an amplification system for simulating radio frequencies.

The basis of the operation is the use of the 35 mm slide projected on the primary screen.

The slide may be a picture of:
1. Single family residence.
2. Apartment structure.
3. Industrial complex.
4. Commercial structure.
5. Target hazard: example lumber yard.
6. Life hazard: example hospital.
7. Water emergency.
8. Vehicle incident: autos, tankers, etc.
9. Aircraft incident.
10. Railroad incident.
11. BELVE.
12. Tank farm.
13. Virtually any type of incident!

The simulation exam will start with only the projected image of the incident location on the screen.

Additional images needed for the incident will be projected over the problem with the use of the overhead projector.

The overhead projector will show red for fire and a light grey for smoke. These images are blocked out by a light shield (usually sand) until they are needed for the problem or fire behavior.

The exam proctor will create a problem by allowing the different colors to come through the light shield. He may create an assortment of visual effects in various locations.

The exam proctor can control the fire behavior to respond to the actions taken by the candidate or as instructed by the exam board. Fires may extend vertically, horizontally, create exposure fires, etc.

The candidate can restrict the fire or smoke spread by the use of his divisions of role players, thus giving the image of fire control.

The candidate and the role players utilize the communication system to simulate communication:

1. Officers.

2. Outside agencies.

3. Dispatch center.

4. ETC.

Some simulator exams may have a separate amplification system to be utilized for background noises of fire, fire engines, sirens, explosions, ETC.

Display boards, sometimes three dimensional if a projected system is used, will display where:

1. Equipment is located.

2. Hose lines are laid.

3. Hose lines are in use.

4. Ventilation takes place.

5. Entry is made. ETC.

ROLE PLAYING:

Each candidate will be assigned ROLE for the exam. His role usually will be as IC (Incident Commander), but may also be:

1. Company Officer.

2. Chief Officer.

3. Dispatcher.

4. Facilitator. ETC.

After the candidate is assigned his role he will be assigned role players for the other needed positions. Each of the role players will convey the responsibilities and functions of their assigned roles. They will react using sanctioned principles of fire ground operations and tactics. The candidate will be required to issue the proper orders and to respond and communicate as he would during an actual incident.

The candidate and the role players will be given a scenario of written and/or oral facts. The candidate and the role players should begin communications, actions and orders based upon their own assessment of the incident situation.

The candidate should use his own experience and knowledge that he has experienced and learned during his career.

The primary companies of the Fire Service are:

1. Engine Companies.

2. Truck and/or Ladder Companies.

3. Salvage Companies.

4. Rescue Companies and/or Paramedic Units.

The basic Firefighting unit of the Fire Service is the ENGINE COMPANY which usually is assigned the functions of:

1. Fire Extinguishment.

2. Exposure protection.

The primary functions of a TRUCK COMPANY are:

1. Ladder operations.

2. Ventilation/Overhaul procedures.

3. Forcible entry.

4. Rescue operations.

5. Utility control.

6. Salvage operations. (unless department has salvage company for these operations)

The primary function of a SALVAGE COMPANY is:

1. Salvage.

The primary function of a RESCUE COMPANY is:

1. On scene emergency medical procedures.

THINGS THAT YOU CAN EXPECT THE EXAM TO CONSIST OF:

1. The use of a representative cross section of typical occupancies.
2. Handout material.
3. Realistic time tables.
4. Results to correspond to actions taken.
5. The use of slides and diagrams of local buildings and areas when possible.
6. Opportunity to familiarize yourself with the equipment.
7. A reasonable amount of information pertaining to the problem.
8. Problem will not build too fast.
9. Simulation area will not be overcrowded with people or equipment.
10. No tricks in the problem.
11. Realistic situations.
12. Problems will not escalate beyond the control of the proctors.
13. Various types of problems.

EXAM FORMAT:

The Fire Simulator Proctors will usually create a situation where you will be the first in officer.

The first in officer is expected to:

1. Take command.
2. Assume responsibility of establishing an incident command post.
3. Directing additional incoming companies.
4. Evaluating personnel.
5. Evaluating needs for apparatus and equipment.
6. Requesting additional manpower, apparatus, and equipment. (including special equipment)
7. Requesting traffic control and/or any other special services required for the incident.

At a structure fire the Fire Simulator Proctors usually will expect you to set the INCIDENT COMMAND POST up in the front of the structure. This will allow you to guide units coming on scene and communicate with radio dispatch. You as the Incident Command officer must remain at the Incident Command Post, therefore you must turn your company over to an acting officer.

SOME OF THE ADVANTAGES OF THE COMMAND POST ARE:

1. A stationary position.
2. Quite place to think and make decisions.
3. A vantage point.
4. Adequate lighting.
5. Area to record.
6. Radios that are more powerful.

INITIAL REPORT:

Immediately upon arrival of the scene of the incident make a SIZE-UP of the situation and report it to dispatch. This report is important to all the responding apparatus.

The description should focus on:

1. Life threatening situations.
2. Location of fire.
3. Travel direction of the fire.
4. Exposure problems.

Report should at least consist of:

1. Your location.
2. Type of incident.
3. Severity of incident.
4. What you need.

EXAMPLE OF AN INITIAL REPORT:

"Engine 82 to dispatch, we have large amounts of fire and smoke coming from the roof of a single story, wood frame, stucco exterior industrial building, about 100 feet wide by about 150 feet deep.

Significant exposure problem on the east side of the building.

Dispatch a 2nd alarm (mutual aid, automatic aid, or just ask for specific units, ie.: three additional engine companies and an additional Aerial Ladder Truck).

We are laying a line to the front of the incident, this will be the location of the Incident Command Post.

Dispatch Police Department for traffic control.

What other units are responding on this call?"

You will be expected to know what your departments initial response will be, but you may not be aware of other units out of service on another call or whatever.

IC OPERATIONS TO BE COMPLETED:

Following is a list of some of the more important operations that the Incident Commander should oversee and follow-up to completion:

1. Life safety and rescue operations.
2. Confinement of the fire.
3. Extinguishment of fire.
4. Exposure protection.
5. Ladder operations.
6. Overhaul.
7. Ventilation.
8. Forcible entry.
9. Salvage.
10. Control of all the utilities.

ESTIMATING MANPOWER:

Fire simulator exam proctors will almost always give you a situation that requires more manpower than you will have in your first in response.
Remember that you are the Incident Commander and that you are expected to know if additional help is needed and how much! you will be expected to justify what you ask for.

A general plan to follow is to ask for the number of companies that you estimate you will need, then add one additional company just to play it safe.
The exam proctors usually will except an excuse for asking for too many companies, but will not except excuses for coming up short.

As the candidate you should have some method of estimating the number and type of companies you will need. This will help you in estimating the manpower needed and also give you a reason to explain to the proctors as to why you handled the situation the way that you did.

Discuss with officers on your Department as to the guidelines that they use for estimating fire companies in different situations.

If additional companies are needed during the exam, it is not your responsibility to determine where they come from. Your are expected to determine what is needed and then request it.

ROUTING RESPONDING COMPANIES:

Fire Simulator Proctors will be watching to see that you follow your Departments policies as far as firefighting procedures and as to where you place your responding companies.

When routing your companies, remember the operations that are to be completed and the primary functions of each company!

UNUSUAL INCIDENTS:

There are many unique circumstances that you may be put into during the exam such as:

1. Hazardous materials incidents.
2. Incidents involving unfamiliar materials.
3. Incidents or circumstances unfamiliar to you.
4. Incidents demanding unique information that you do not possess.

In handling these types of incidents you can follow some of these suggestions:

1. Do not do anything that may place personnel or the public in danger.
2. Do not do anything to worsen the incident.
3. Acquire as much information regarding the incident as possible before taking any course of action.
4. Get advice from authorities that specialize in whatever the incident involves. Remember that nobody knows everything and that it is not a discredit to seek help.
5. Also be careful not to overestimate the incident.

THINGS TO KNOW WHEN COMMANDING AN INCIDENT:

1. Where to assign your company.
2. Where to set-up the Incident Command Post.
3. Priorities of incident.
4. Number of engine companies needed for fire extinguishment and exposure protection.
5. How many truck, salvage, and rescue companies you will need.
6. What special apparatus and equipment will you need.
7. The initial responding units.

8. Where to assign the first and second in engine companies. (if needed)

9. Where you intend to place other incoming engine companies.(if needed)

10. What task to assign the first and second in truck companies. (if needed)

11. Do you have an sufficient amount of units to handle all operations.

12. Your initial report to dispatch.

13. Orders for all companies that are responding.

14. If special services are needed, ie.: Utility Companies, or Police Department for traffic control, etc.

POST EXAM:

Sometimes after the simulator portion of the exam, the exam proctors may call the candidate back into the exam area and ask him to explain his:

1. Responses and actions.

2. Decisions or theories, why they were appropriate.

3. ETC.

HOW SIMULATOR EXAMS ARE GRADED:

Some things to keep in mind when taking the simulator portion of the promotional exam:

1. Remember the elements of communication.

2. Have broad-minded objectives and give orders in those terms.

3. Presume subordinate officers will assume and respond to orders given in broad terms.

4. Recognize that strategy is an overall blueprint by which the crisis is handled.

5. Recognize that tactics is the attempt that is directed toward the accomplishment of the plan.

6. Sustain a concise thought process and relinquish the particulars to the Firefighters.

7. Remember large fires are somewhat like small ones, however they need more men and equipment and take more time to extinguish.

8. Remember that large fires require more fire flow and that it must be applied at greater distances.

9. Delegate control criteria based on objectives. These objectives should have the measure of quantity, quality and time.

10. Recommend that companies stay together in the sector assigned.

11. Encourage company officers to use judgement and assume a command role if need be.

12. Remember the particular objectives that may need to be met:
    Rescue and evacuation of inhabitants
    Life safety of Firefighters
    Protection of exposures
    Confinement of the fire
    Fire extinguishment
    Overhaul of the fire scene
    Ventilation
    Salvage

13. Avoid TUNNEL VISION in directing operations.

14. Factors for establishing objectives:
    Construction of structure
    Size of fire
    Direction of fire travel
    Occupancy classification
    Safety of Firefighters
    Safety of Citizens
    Potential of the emergency
    Priorities of the operation
    Weather
    traffic
    water supply
    Manpower
    Equipment available
    Resources available

15. Remember to set up a command post.

16. Inform all subordinate officers aware of command post location.

17. Initiate a strong command structure and maintain unity of command.

18. If needed redeploy companies.

19. Recognize reflex times in handling the emergency.

FIRE SIMULATOR EXAM GRADING SHEET:

During the Simulator Exam the Proctors will have a grading sheet in front of them. This sheet will consist of a list containing the components that should be used by the IC during an incident.

These components will usually be rated as:

1. Exceptional or outstanding.
2. Good or above average.
3. Average.
4. Below average or poor.

Sometimes the raters may use a number scale such as 1 to 10 or 1 to 15, with the highest number being outstanding and the lowest number being below average or poor. The rating sheets will almost always have aa area for additional comments.

The grading sheet will usually consist of the following components:

1. Size-Up, Initial report, 952 report:
   Evaluation of the incident.
   Clarity of orders.
   Brief with an objective.
   Notification for additional manpower.

2. Strategy and Tactics:
   Use of manpower and equipment. (on scene and incoming units and officers)
   Location of hoselines.
   Internal and external exposures covered.
   Ventilation.

3. Control of:
   Self and the problem.

4. Were objectives based on:
   Time.
   Quality.
   Quantity.

5. Productive use of order with all incoming units: With particular emergency incident.

6. Effective fire ground organization using:
   Chain of control, chain of command.
   Span of control.
   Unity of control.
   Division of manpower.
   Staging areas.
   Operations of command post.
   Coordination of decisions. (decision loop)

7. Effective communication techniques with other personnel.

8. Use of open-minded objectives, while exercising good management techniques.

EXAMPLE OF A SIMULATOR EXAM INCIDENT:

REMEMBER:

1. Control incident. (give the situation direction)

2. Get "Feed Back".

3. One assignment to each company at a time.

4. Give direction not details.

5. Use equipment and manpower as needed, save the balance until in staging needed.

6. Log events as they occur.

7. Keep command available: delegate responsibilities.

8. Use special advisors as needed, if available.

9. Be aware of communications.

10. As IC, coordination is a primary goal.

11. During actual simulator exam you will have drawings, pictures, slides, and an amplifying system (audio and visual effects). This will assist you in your initial report and keep you apprised of situation as to what effects you are having on the incident. Don't forget to watch and listen to these devices so that you can react to the changes that are effected by your delegated assignments.

Assume that you are the first in officer for the following incident. You will be responding with your Departments normal response of equipment and manpower with the exception that your Departments normal Incident Commander will not be responding.
(unless otherwise stated)

In this exercises direction will be determined by the use of the terms; "Front, rear, left, or right of the structure". The normal terms of north, south, east, and west bearings will be of no value in these descriptions. (in an actual simulator exam it is possible that you will use either of the terms because of the visual equipment)

KEEP IN MIND THAT IN THE FOLLOWING EXAMPLE, THAT IT IS ONLY AN EXAMPLE AND NOT NECESSARILY THE BEST OR ONLY WAY TO MANAGE THIS INCIDENT. THE EXAMPLE IS PRESENTED AS A METHOD TO ENCOURAGE YOU TO LOGICALLY APPRAISE THE INCIDENT FROM BEGINNING TO CONCLUSION. THE MORE THAT YOU PRACTICE COMMANDING AN INCIDENT, THE MORE CONFIDENT YOU WILL BECOME WITH EACH INCIDENT. GO THROUGH THE INCIDENT EXAMPLE SEVERAL TIMES AND ADD DIFFERENT CONDITIONS
EACH TIME. CREATE INCIDENTS THAT BECOME MORE COMPLICATED AND SEE HOW YOU ADJUST. MANAGE EACH INCIDENT ACCORDING TO YOUR PARTICULAR FIRE DEPARTMENTS STANDARD OPERATING PROCEDURES, RESPONSE LEVELS ETC!

EXAMPLE OF SIMULATOR INCIDENT

DESCRIPTION OF INCIDENT:
FIRE SINGLE FAMILY RESIDENCE

It is 0100 hrs on a Monday morning, no wind, clouds or precipitation.

You have a single story, single family dwelling, possibly occupied with two adults and two small children.

Upon arrival there is smoke and flames coming out of the kitchen area on the right side of the dwelling.

There are no exposure problems.

Special problem = upon arrival the owner of the home states to you that all of his family is accounted for except for his wife.

First alarm response = E-81, T-81, R-81
B/C-81 is unable to respond.

THINGS TO DO DURING THIS INCIDENT:

1. Call for additional help.
2. Set up command post and give its location.
3. Call P.D. for crowd and traffic control.
4. Initiate search and rescue.
5. Ventilate.
6. Extinguish the fire.
7. Command assignments.
8. Cover vacant Fire Station with manpower and equipment.
9. Use proper timing of manpower and equipment.
10. Check for fire extension.
11. Manage utilities.
12. Forcible entry.
13. Overhaul and salvage.

POSSIBLE INITIAL REPORT:

"Dispatch, E-81 is on the scene of a occupied one story single family residence with fire and smoke showing from the right side of the structure.

We have laid a line to the front of the structure, this will be the location of the Incident Command Post.

E-81 will conduct search and rescue.

Dispatch appropriate units to cover vacated stations.

Dispatch P.D. for traffic control."

POSSIBLE FOLLOW-UP ASSIGNMENTS:

T-81 will ventilate the structure.

R-81 will relieve E-81 in search and rescue.

E-81 will attack the fire from inside structure, after ventilation is completed.

T-81 could manage utilities after ventilation is completed.

R-81 could check for extension of fire.

Continue to communicate with all units for updated reports of information and react appropriately until the incident is concluded.

This incident could escalate or the fire could be extinguished!

For a complete guide to fire Assessment Centers and simulator exams see: **"FIRE CAPTAIN ORAL EXAM STUDY GUIDE"**, available from: INFORMATION GUIDES.

SECTION #4

AFTER PROMOTING TO THE POSITION OF FIRE CAPTAIN

AS FIRE CAPTAIN

Be aware and perform the **RESPONSIBILITIES** and duties for the position of Fire Captain:

1. Management, supervision.
2. Organization.
3. Planning.
4. Morale.
5. Commanding.
6. Orders.
7. Delegation.
8. Training.
9. Firefighting.
10. Fire Prevention.
11. Target Hazards.
12. Water Supply.
13. City geography.
14. Reports, Records.
15. Department policy, procedures, rules and regulations.
16. Budgeting.
17. Arson investigation.
18. Apparatus and equipment.
19. Subordinates.
20. Tours, Demonstrations.
21. Emergency situations.
22. Quarterly task.

Continue to follow and complete the previous three sections: **JOB PERFORMANCE**, intensify your **TRAINING**, **KNOWLEDGE**, and **EXPERIENCE** in Fire Department Administration.

SECTION #5

EXAM ELEMENTS & STRATEGY

EXAM STRATEGY / PRINCIPLES

Many promotional candidates do not promote because they do not know how to prepare for the promotional process and they do not know how to take exams.

It is not just a matter of study, its knowing what and how to study, along with how to take exams, including the proper management of time.

Furthermore, taking test is not just a matter of answering the questions that the candidate may be sure of the answers, candidates need to know strategy and deduction in answering the questions that they do not know the answers for.

Candidates should be aware that most Fire Departments will promote the candidates with the highest scores first.

ELEMENTS PERTAINING TO PROMOTIONAL EXAM PROCTORS:

The exam proctors, usually a person from the City Personnel Department, is going to want to comprise good questions, that will have several possible answers, in order to challenge the candidates to reason, even though there will/should be only one correct answer.

The exam proctor may make use of personnel from the Fire Department, including Fire Departments from surrounding jurisdictions, when preparing the exam. The proctor may also get exam information from prior exams or even from professional exam producing companies.

The proctor will construct the exam from the above mentioned sources along with books that relate to the position that is being tested for, including subjects such as:

1. General knowledge of firefighting.
2. Department rules and regulations.
3. Local building codes.
4. Local fire codes.
5. State codes.
6. Department manuals.
7. Any subject matter pertaining to the particular position being tested for.

After the proctor determines the subject matter of the exam, he will decide as to what form he will use for the questions.

The questions in a promotional exam will be presented in the following **FORMATS:**

1. Multiple choice.
2. True and false.
3. Fill in the blanks.
4. Essay.
5. Matching.

The subject matter included in promotional exams is usually presented in the form of the "MULTIPLE-CHOICE EXAM".

MULTIPLE CHOICE EXAMS require that you choose an appropriate answer from a list of answers that have been offered for a question that has been presented. (usually three to five choices)

The candidate is to choose the one answer that is the best one, the one that most nearly or most often is correct.

Sometimes there is no answer that is complete, or exactly correct, or always correct.

Choose the best answer. The best answer is the one that is most nearly right, under ordinary conditions.

The candidate must be sure he/she knows what the question asks and what the choices say.

On every test, candidates choose wrong answers simply because they failed to read the question carefully, or failed to pay attention to a part of the question.

EXAMPLE:

The number of days in a year is:
A. 365.
B. 366
C. 367
D. 368.

The answer you should choose is choice A" because it is the one which is most often correct. Choice "B" is true for leap years, but most years have 365 days. Choice "A" is the best answer.

TRUE FALSE QUESTIONS

True and false questions are usually statements that are true or false. You must decide if the statement is accurate (true) or inaccurate (false).

EXAMPLE:

1. When water freezes it expands. (T or F)

   Answer = T.

2. The term used to express electrical pressure is amperage. (T or F)

   Answer = F, correct term is volt.

3. The term which is applied to the distance from the center of a circle to its exterior is radius. (T or F)

   Answer = T.

4. An electric fuse is used to reduce current. (T or F)

   Answer = F, correct use is to protect against electrical overload.

5. The method by which the sun transmits heat to the earth is known as radiation. (T or F)

   Answer = T.

ESSAY TEST

In an essay test you will be compelled to respond in a written account to a question or statement. (this type of exam is rarely used for entrance level Firefighter exams)

EXAMPLE

1. Describe SIZE-UP.

   Answer = SIZE-UP is the mental evaluation made by the officer in charge which enables him to determine a course of action. SIZE-UP includes such factors as time, location, nature of occurrence, life hazard, exposure, property involved, nature and extent of fire, available water supply and other firefighting facilities. This report is usually given via radio.

2. Describe WATER, include how it applies to the Fire Service.

   Answer = WATER, H2O is a colorless, odorless, tasteless liquid used widespread for extinguishing fire. Its forms may be liquid, ice (solid), or steam (vapor). The freezing temperature of water is 32 degrees F. The boiling point of water is 212 degrees F. WATER is effective as a cooling agent because it has the largest "latent heat" of vaporization of all common materials, 90 Btu per pound.

3. Describe CORROSIVE CHEMICALS.

   Answer = CORROSIVE CHEMICALS are chemicals that are particularly injurious to life as, solids, liquids, or gaseous substances they burn, irritate, or destroy organic tissue, most notably the skin and, when taken internally, the lungs and stomach. Include acids, anhydrides, and alkalis.

4. There are three types of radioactive particles, name and rate them in order of danger.

   Answer = The three types of radioactive particles and there danger levels are: Alpha, these particles can be stopped by the outer layer of skin. Beta, these particles might penetrate to a third of an inch before being absorbed, protective clothing will usually give protection. Gamma, these are the worst ones, they can go through the body like X-Rays and cause serious injury.

5. What is the method for using a pressure water Fire Extinguisher?

   Answer = The pressure water Fire Extinguisher is most effective when applied close to the fire. Pressure water Fire Extinguisher can be directed from distances of from 30 feet to 40 feet horizontally. It is recommended that the operator should use a series of short rapid strokes to produce a continuous stream; keep the nozzle as close to the fire as possible; direct the stream to the base of the fire; and use spray or straight stream.

MATCHING QUESTIONS

With matching questions, you will be given two columns of information, each column will have certain details, dates, facts, or a statement. The objective is to accurately merge items from each column together.

EXAMPLE

1. Match the items in column I with the correct items in column II.

   I
   A. Rungs
   B. Screw
   C. Nail
   D. Nut
   E. Hex head bolt

   II
   1. Screw driver
   2. Hammer
   3. Ladder
   4. Allen wrench
   5. Crescent wrench

   Answers = A and 3; B and 1; C and 2; D and 5; E and 4.

2. Match the correct fire class rating in column 1 with the correct color and shape designation in column 2.

   1
   A. Class A fires
   B. Class B fires
   C. Class C fires
   D. Class D fires

   2
   1. Red square
   2. Yellow star
   3. Blue circle
   4. Green triangle.

   Answers = A and 4; B and 1; C and 3; D and 2.

3. Match the correct fire class rating in column 1 with the correct type of fire in column 2.

   1
   A. Class A fires
   B. Class B fires
   C. Class C fires
   D. Class D fires

   2
   1. Electrical fires
   2. Metal fires
   3. Ordinary combustibles
   4. Flammable liquids

   Answers = A and 3; B and 4; C and 1; D and 2.

4. Match the occupancies in column I with the proper occupancy classification in column II.

I	II
A. Educational	1. A occupancy
B. Hazardous	2. B occupancy
C. Assembly	3. E occupancy
D. Business	4. H occupancy
E. Institution	5. I occupancy
F. Carports/fences	6. R occupancy
G. Residences	7. M occupancy

Answers = A and 3; B and 4; C and 1; D and 2; E and 1; F and 7; G and 6.

5. Match the proper terms in column I with the corresponding type of tire wear in column II.

I	II
A. Underinflated	1. Tire wear on outside
B. Overinflated	2. Feathering
C. Out of line	3. Flat spots/abrasions
D. Improper balance	4. Flat spots/cupping
E. Heavy braking	5. Wear on one edge
F. Toe in or out	6. Center of tread wear
G. Camber	7. Both edges wear

Answers = A and 7; B and 6; C and 5; D and 4; E and 3; F and 2; G and 1.

FILL IN THE BLANKS QUESTIONS

With fill in the blanks questions, it is necessary for you to fill in a missing word/words so as to accurately complete a statement. These type of questions eliminates guessing from suggested answers, summon the correct answer from memory.

EXAMPLE

1. The freezing point on a Fahrenheit thermometer is _____ degrees.

   Answer = 32.

2. A molecule, is the term which is generally applied to the smallest units into which a substance may be _____ without destroying it.

   Answer = divided.

3. Materials which offer a very high or very great resistance to the passage of electricity are known as an _____.

   Answer = insulators.

4. To prevent the passage of gas or air into plumbing fixtures a _____ is used.

   Answer = trap.

5. If substance is soluble in water, it will ____.

   Answer = dissolve.

QUESTION TYPES

After the proctor determines the subject matter and the format of the exam, the proctor will have to decide whether the questions will:

1. Factual type questions; there is no room for discussion on these type of questions, either the candidate knows the answer or not.

2. Opinion/statement, based on a reference, the answers to these type of questions are based on opinions, which may be open to discussion.

3. Opinion/statement, based on translation; with these type of questions the answer is based on interpretation, such as an incorrect word or phrase.

4. Data based on general firefighting principles; these type of questions are based on an accepted principle.

5. Job Announcement sheet/flyer; these questions are based on information included on the positions Job Announcement flyer.

JOB ANNOUNCEMENTS

The proctor/examining authority will have to prepare the Job Announcement. (candidates should memorize all the data included on this flyer)

EXAMPLES OF TYPICAL FORMAT AND THE INFORMATION CONTAINED IN PROMOTIONAL JOB ANNOUNCEMENTS/FLYERS ARE ON THE FOLLOWING TWO PAGES

"THIS CITY" OFFERS EMPLOYMENT AS: **FIRE ENGINEER**

THE POSITION, MONTHLY SALARY = $3,500.00

Under general supervision - drives standard fire engines, aerial platform (ladder truck), and other automotive equipment in responding to fire and emergency calls; operates fire pumps and other apparatus in the suppression of fires; assist in inspecting and pre-fire planning inspections; and does related work as required and assigned. This position reports directly to a Fire Captain, and is primarily responsible for the proper driving and operation of equipment assigned.

Thorough knowledge of principles of hydraulics applied to fire suppression; of engine and ladder truck driving and operation; knowledge of technical fire fighting techniques; modern fire suppression methods; of departmental rules\regulations, first aid and the operation of resuscitators, geography of the City and location of water mains/hydrants.

QUALIFICATIONS

EDUCATION AND EXPERIENCE: Three (3) years of full-time experience as a Fire Fighter with a full-time professional department and completion of 18 units at a accredited college, of which a minimum of 6 units shall consist of hydraulics and automotive apparatus and equipment. (CLOSED PROMOTIONAL)
LICENSE: Valid State Driver's License.

THE EXAMINATION

EXAMINATION PROCEDURE: A written exam will be given to all candidates who qualify for the position on : 3-3-91, at 9:00AM in the Headquarters Classroom. Those passing the written exam will be given a Practical Exam to test apparatus and driving skills and pump operation knowledge on 3-9-91.

Candidates passing both written and practical exams will then be invited to participate in an Oral Interview, and an In-House Evaluation, on 3-12-91.

Pass or fail notification, for each phase, will be sent to the Training Officer.

Candidates must successfully complete all phases of the exam process, with a minimum score of 70% to be placed on the Eligibility List.

EXAM WEIGHTS

PRACTICAL	= 30%	WRITTEN	= 25%
ORAL INTERVIEW	= 20%	IN HOUSE EVALUATION	= 25%

FINAL FILING DATE: 2-20-91. POSTMARKS NOT ACCEPTED.

NOTE: APPLICATION MATERIAL MUST BE SUBMITTED IN TRIPLICATE AND TRANSCRIPTS INDICATING COMPLETED UNITS MUST BE FILED WITH APPLICATION.

"THIS CITY" OFFERS EMPLOYMENT AS: **FIRE CAPTAIN**

THE POSITION, MONTHLY SALARY = $4,000.00

Under direction, commands a fire company and fire station on an assigned shift, responding to fires and emergencies; supervises a systematic program of company fire prevention and pre-fire planning inspections; exercises leadership in instructing personnel, applying policies and procedures, and in performing related work as required and assigned. This position reports directly to the Battalion Chief.

Must have thorough knowledge of the principles of fire prevention and suppression; of the use and maintenance of firefighting equipment and apparatus; of pre-fire planning inspection methods; good knowledge of the principles of supervision and training, with leadership and supervisory ability; of mutual aid and cooperative response areas; of hazardous materials and communications. Must have the ability to prepare reports and maintain records of inspection.

QUALIFICATIONS

EDUCATION AND EXPERIENCE: Five years full-time experience as a Firefighter with this Fire Department; and completion of 30 units at an accredited college. (CLOSED PROMOTIONAL)
LICENSE: Valid State's Driver's License.

THE EXAMINATION

EXAMINATION PROCEDURE: Assessment Center consisting of individual and group exercises designed to measure ability to perform the duties of Fire Captain.

The written exam is scheduled for 4-2-91, at 9:00AM in the Headquarters Classroom. Those successful in the written portion will then be invited to participate in a Fire Simulator exercise, scheduled for 4-9-91, at 9:00AM in the Headquarters Classroom. The Assessment Center will begin 4-12-91, at 9:00AM, followed by an In-House Evaluation. Candidates must pass the Written and Simulator Exams to be eligible for the Assessment Center.

EXAM WEIGHTS

WRITTEN EXAM = PASS/FAIL   SIMULATOR EXAM = 25%
ASSESSMENT CENTER = 50%   IN-HOUSE EVALUATION = 25%

FINAL FILING DATE 3-20-91, POSTMARKS NOT ACCEPTED.

NOTE: APPLICATION MATERIAL MUST BE SUBMITTED AS FOLLOWS - ORIGINALS PLUS THREE (3) COPIES AND TRANSCRIPTS INDICATING COMPLETED UNITS MUST BE FILED WITH APPLICATION. ALL NOTIFICATIONS WILL BE SENT TO THE DEPUTY CHIEF.

AN EQUAL OPPORTUNITY EMPLOYER

EXAM CANDIDATES

The promotional candidate should prepare himself/herself for the particular position he/she is going to test for first, by following the suggestions included in the previous chapters of this book.

Candidates should break down the Job Announcement and compile a list of what is expected for the position and make sure that he/she qualifies for all of these requirements.

Remember proctors may base some exam questions on the qualifications and requirements of the position that are listed on the Job Announcement. The candidate should also remember that the proctor is not a Fire Department expert and will use books and reference material to get this information.

Candidates should compile a personal library of Fire Service related books and reference material. There are many books and material that should be included, some of the books are very large, such as N.F.P.A. Handbook.

The candidate may think it impossible to absorb all of the information available, but it is not necessary to absorb all of it!

The candidate should make a list of the books/material that are appropriate for the particular exam and future exams and eliminate the others. Some books may list information that the candidate would never be expected to memorize or retain, such as the numerous hazardous chemicals, etc.

However the number of books/material that a candidate should read or accumulate should not have a limit. Remember the candidate is committed for a lifetime in this career, and is determined to promote.

BASIC EXAM PREPARATION STRATEGY

1. TIME: It is possible to pass exams without taking the time to study/prepare, but it is not likely that a candidate will score high without the proper preparation. Even if the candidate has just missed taking an exam and the next exam is not for another year or two or even three or four years, take the time to study now!

2. DISCUSSIONS: It is a good study technique to exchange information, questions, and answers with fellow candidates in preparation for upcoming promotional exams. Just remember that sometimes it works out to the advantage of the candidate that may not be as astute as others. A candidate that may not be a good studier but is a good listener, stands to gain a lot of information, while not contributing a great deal. On the other hand the candidate that knows a lot of information may give out much more than he/she may receive. Group study is "OK", but remember that each candidate is a competitor! If a candidate knows information that the others do not know, it is probably best not to share this information.

3. STUDYING ON THE MOVE: There are many things that you can study on the go. Candidates can carry notes of particular information that they need/want to study while on the move. Also candidates can take along with them various manuals or STUDY GUIDES such as those that are available from INFORMATION GUIDES. Candidates most likely will not carry larger books around, so it is best to set aside certain times for study in the same place that the candidate associates with good concentration.

4. PRIOR PROMOTIONAL EXAMS: Former promotional exam questions may or may not be available, but if it is possible, the candidate should try to build up a collection of prior exam questions. Sometimes questions can be obtained from fellow workers that have already promoted. Candidates that have taken prior promotional exams should always make a list of all the questions that they remember from the exam. Candidates should compile a list of all the information that they come in contact with!

5. LEGISLATION: Fire and building codes must be learned exactly the way that they are written. The candidate may be able to verbalize the a principle in the area of general firefighting, but with codes each word and comma may be very important. These codes should be memorized.

6. FLASH CARDS: Candidates should prepare flash cards with questions on one side and the answer on the other side. By preparing just a few cards a day candidates will compile a large stack of information. As cards are learned, set them aside, but every once in a while go over all of the flash cards.

7. FIRE DEPARTMENT ADMINISTRATION/SUPERVISION: Most promotional exams within the Fire Service will have questions pertaining to these subjects. It is essential for the candidate of a position or responsibility to be familiar with the principles involved.

BASIC PRINCIPLES OF PROMOTIONAL WRITTEN EXAMS

Some suggestions to follow when taking the written portion of the promotional exam include:

1. Read every word of the job announcement.

2. Analyze the job announcement.

3. List the subjects on the job announcement.

4. Go to test area prepared.

5. Try not to leave the test area for any reason.

6. Listen carefully to directions.

7. Do not ask the proctor unnecessary questions.

8. Try to manage your time during the exam.

9. Answer the easy questions first, so as to get as many correct answers down as you can, keep track of the ones that you have omitted.

10. Stay the full time of the exam; it is better to go over your exam for accidental errors or for further study of some of the more difficult questions.

11. Reason your way to the correct answers.

12. Use care with answers that use the words: always or never.

13. If one of the answers includes another answer the probability is that the broader answer is the correct answer.

14. Answers that simply repeat the question should be eliminated.

15. Answers that contradict the question should be eliminated.

16. When answering situation type questions, try to imagine a specific example of the situation so as to test the possible answers.

17. For questions relating to job duty, remember that public duty comes first.

18. Eliminate answers that imply unworthy characteristics in department personnel.

19. Be careful of questions that imply that books are incorrect.

20. Eliminate answers that you feel are incorrect, then choose the answer that comprises a broad, clear statement of principle and/or the answer that you understand.

21. If the exam does not have a system for deducting wrong answers from correct answers, then guess on the questions that you do not know the answers for, or if you are running out of time. Do not leave any questions unanswered!

22. Use the supplies given to you for taking the exam; such as pencils. If you need to change an answer, erase thoroughly.

The following is a corresponding list/supplement to the above list that should be followed when taking the written portion of the promotional exam:

1. Be cooperative.

2. Follow directions.

3. Make note of your exam test number.

4. Read the whole question carefully; be sure that you know what the question asks and what the choices say.

5. Choose the answer that is generally correct; answer according to what is generally or usually true, not by what would be true in some particular case.

6. Use your time efficiently; Promotional exams are not speed test, but usually will not give you all the time that you might like to have.

7. Make decisions; your decision should be one of the following:

   A. If you know the answer. (answer this question now)

   B. That you can figure out the answer, but that it will take a lot of time. (skip this question and come back to it later)

   C. That you do not know the answer and that you cannot figure it out. (make a guess and answer this question now, unless you have been instructed not to do so)

8. Don't give up; hang in there and give it a full effort.

9. Try not to change too many of your answers; remember that the best answer is the one that is usually or generally right, although if time permits review your exam.

10. Be at your best the day of the exam; be well rested, allow plenty of time to get to the exam, get there early.

11. Return your exam and exam materials to the exam proctor.

AFTER THE EXAM

Upon completion of the exam make sure that you hand in all of the test matter to the proctor.

After the exam you may ask the proctor as to when you will be notified of the exam results.

After you leave the test location, you should go somewhere that you are comfortable and take the time to list of:

1. The type of questions that were on the exam, ie: multiple choice, true or false, matching, fill-in, etc.

2. The areas that were tested.

3. Any question/questions that you feel that you should save on a continual ever-growing list of test questions for your future reference.

4. Anything or feeling that you have about this particular exam that you might want to recall at a later date.

5. Any ideas that you may have on how you could have improved your preparation for this exam.

BASIC PRINCIPLES OF PROMOTIONAL ORAL EXAMS
WHY THEY HAVE ORAL EXAMS

The purpose of an ORAL INTERVIEW is to evaluate each candidates personal qualification, training, experience, attitude, personal attributes, and any intangibles involved that usually outline a candidates likelihood of success or failure for the promotional position.

The objective of the ORAL INTERVIEW is to determine and identify various elements that have not been examined in the other phases of the exam process. Some examples are:

1. Adaptability.

2. Attitude.

3. Ability to express yourself.

4. Ability to function under stressful situations.

5. Ability to work with others.

6. Ability to follow directions.

7. Your judgement.
8. Your attitude to working required hours.
9. Poise.
10. Stability.
11. Your personal grooming habits.
12. Integrity.
13. Enthusiasm.
14. ETC.

ORAL INTERVIEW

The ORAL INTERVIEW is almost never aimed at testing your professional knowledge. This has, hopefully, previously been done or will be done in the written examination, in the checking of your application/resume, in reviewing your work record, in reports from your employers, superiors, and associates.

Remember that you asked for this opportunity to be interviewed when you filled for your application. You are not here against your will, and can choose to withdraw at any time.

The decision to be or not to be interviewed is yours to make. You are being interviewed because thus far you have shown the basic qualifications, technical and/or intellectual abilities, along with the experience being sought. You are still in the running and this is an important step in the selection process.

HAVE A POSITIVE ATTITUDE CONCERNING THE ORAL INTERVIEW, BE ENCOURAGED THAT YOU ARE STILL IN THE RUNNING!

TYPES OF ORAL BOARDS

Fire Department oral boards usually consist of three members that may be selected from:

1. The community.
2. Business and industry.
3. Fire Department members.
4. Members from other Fire Departments.
5. Minority group representatives.
6. Personnel Department.

ORAL BOARDS

Oral interview boards are set up in different areas in different ways and are composed of various groups of interviewers. In some cases the oral boards are composed entirely of lay members of the Department for which the test is given. Some interview boards are composed of members from other Fire Departments. There is no set standard or criteria as to where, who, how many, or from what positions the board may be chosen from, although certain standards are attempted.

ORAL INTERVIEWS may be one of two types:

1. Stressful.
2. Non-Stressful.

STRESSFUL ORAL BOARDS will attempt to evaluate you by creating a stressful atmosphere so as to see how you will perform under this stress. In this interview without introduction you will be asked a question which requires taking a stand on a controversial subject, and then all of the board members will take the opposite view point. The chairman of the board or one of the members previously designated by the board, will inject new controversies on different subjects at random intervals, never letting you fully explain your position on any subject. These types of interviews are seldom used.

NON-STRESSFUL ORAL BOARDS (ordinary interview) are normally of a cordial atmosphere where the board will attempt to put you at ease during the complete interview. This interview usually starts with some type of introductory remark that is designed to reassure you, and is followed by some review questions that you are familiar with. These types of questions are designed to show your interest, vocabulary, along with the volume and tone of your voice, also your mannerisms, etc.

HOW TO PREPARE FOR THE ORAL INTERVIEW

You have been preparing for the ORAL INTERVIEW portion of the exam from the time that you started to learn about the Fire Service as a career.

Some of the PRE-ORAL INTERVIEW preparation that you have already established include:

1. Submission of your application and resume will have informed the oral board as to your sincerity and desire to obtain the position, along with showing your ability to organize your thoughts, your neatness, and your background. This is the first impression that the oral board will have of you, make it good!

2. Knowledge gained concerning the position.

3. Knowledge gained concerning the Department.

4. Knowledge gained concerning the community.

5. Education and training.

6. Knowing your competition.

7. Knowing who is getting promoted.

8. Self assessment.

9. Your exam checklist.

10. Practice exams/interviews.

ORAL INTERVIEW EXAM CHECKLIST

1. Know what's on your application and resume.

2. Know the duties/responsibilities of the position.

3. Know the qualifications of the position.

4. Know the Department.

5. Be prepared to present yourself effectively.

6. Know what you have done and can do.

7. know what the job requires in the way of performance.

8. Visualize questions/answers of likely questions.

9. Visualize answers to questions.

10. **PRACTICE.**

THE ORAL INTERVIEW

The ORAL INTERVIEW for Fire Department promotional exams will last any where from 20 to 30 minutes to as much as an hour.

On the day of the ORAL INTERVIEW:

1. Arrive early, know where you are going, where you are to park in advance, allow time to get there, park, and reach the interview room, announce your arrival to the proper person, find a place to sit and relax until you are called.
2. Bring your application for review, or some light reading.
3. Do not bring any exhibits or technical material unless you have been instructed to do so.
4. Be clean and well groomed.
5. Be neat.
6. Avoid all extremes in dress and hair style.
7. Do not wear any pins or emblems.
8. Suit, sport-coat and tie are recommended.
9. While waiting to be called, recheck your attire
10. While waiting restudy your application.

When you are called for the interview, remember that the interview is a sales interview and that you are the salesman and that the board is the client, and that the product that you are selling is you.

Upon entering the interview room, politely acknowledge any introductions which may be made before you sit down.

If you know a member of the board do not try to hide it, but you do not have to emphasize it either.

Usually the interview board will be seated on one side of a table and you will see an empty chair for you on the opposite side of the table, but there is no set policy regarding the seating arrangements.

Enter the room standing erect and walk with confidence.

You will be introduced to each member of the board, as you are introduced, look each in the eyes and acknowledge along with a firm hand shake if the opportunity is extended. Be confident and direct, use rank titles if applicable.

If the board members are introduced by name, try to remember the names and where possible address the members by name during the interview and/or at its conclusion. Sirs will be satisfactory if you think that you may become confused. Do not confuse title or rank between board members.

You will now be asked to take a seat, sit erect in the chair, do not seem stiff and rigid, be erect and comfortable. Don't shift positions in your seat constantly.

It is alright to gesture with your hands, but not too much! You do not want the board staring at your hands. Do not sit on your hands, put your hands in your lap or on the table in front of you. Be natural in your expression and movements. Avoid distracting movements, such as scratching, pulling on buttons etc.

When asked questions look at the face of the person asking the question as this will help you to focus on the question.

When answering a question, direct the answer to the person that asked the question, but do not ignore the other members of the board.

The board will usually review your application at the start of the interview. Do not interrupt unless there is a significant error made, do not quibble over matters of minor importance.

After the board reviews your application, the first question usually will be a question such as: Tell us something about yourself or Why do you want to be a Fireman? Try to have some opening statement prepared about yourself, your background and qualification and why you want to be a Fireman. Be careful here, never memorize and recite a prepared statement, just have an organized list in your head that you would feel comfortable elaborating on.

Do not give one word answers to questions, try to let the board know why an answer is "yes" or "no", sell yourself!

Answer questions confidently and to the best of your ability. Do not try to "fake it", be honest, if you do not know an answer to a question tell the board that you do not know the answer. If you lie, you can be disqualified! Be sincere and deliberate with your answers.

Remember that you want this position, sell yourself to the board, don't sell yourself short. The board members are trying to select the best man for the position, you must give them something to work with.

Don't worry if you are nervous, the board expects this. Most board members usually will want you to do your best. The board members are not your enemies, they are just observers looking for the best candidate for the position. Let the board know that you are the best candidate for the position.

Let the board know that you are a serious candidate, act professional at all times during the interview. If appropriate a little humor is alright, but do not overdo it. Be attentive.

When answering questions:

1. Modulate your voice, don't speak in a monotone manner.

2. Speak-up, but not too loudly.

3. Make sure that you understand the question before answering, if not, restate the question or ask for a clarification. Don't overdue this, do not make the board repeat every question.

4. Be honest.

5. Be pleasant, smile occasionally. Don't' be stone faced.

6. Wait until the entire question is asked, don't interrupt the questions.

7. Do not argue. If you make a stand on an issue, do not change your opinion unless you are proven wrong.

8. Answer questions completely and then stop, don't try to dominate the interview.

9. Expect abrupt changes in the questioning.

10. Don't wisecrack.

11. Reply with your answers promptly, not hastily.

12. Don't spend too much time praising your current job, you are after all looking for a new career.

Sometimes a board member may stop you in the middle of answering a question, this could be a positive indication that he feels that you have already answered the question to his liking and he wants to cover another area. Never try to continue to answer the question! Be ready for the next question.

Just prior to the end of the interview, the board will usually ask if there is anything that you would like to add:

1. If they have overlooked one or more of your strong points, you should bring it out now.

2. Do not start an extended presentation at this point.

3. If you have nothing to add, just say; No thank you, I believe we have covered everything.

4. Don't compliment the board members.

The board will let you know when the interview is over, at this time just thank them, shake their hands and leave the room. Do not make a speech at this time. The only time that you should say anything is if something very important has been left out, then you should briefly mention it. You actually terminate the interview, this is the point where you can actually talk yourself out of a good impression or fail to present an important bit of information.

When you leave the interview room you should leave with poise and confidence watch for an offer of a handshake, if an offer is made, shake hands.

HOW TO PUT YOUR BEST FOOT FORWARD

Throughout this process, you may feel that the oral board is trying to penetrate your defenses, to seek out your hidden weaknesses, and to try to embarrass or confuse you. This is not the case, the are compelled to make an evaluation of your qualifications for the position. They want to see you at your best.

The oral board must interview each candidate and a noncooperative candidate may become a failure in spite of the boards best efforts to bring out the candidates qualifications. You must put your best foot forward, the following are some suggestions on how to accomplish this task:

BE NATURAL: keep your attitude confident, but not cocky. If you are not confident that you can do the job, don't expect the board members to be. Don't apologize for your weaknesses, bring your strong points. The board is interested in a positive, not a negative presentation. Cockiness will antagonize any board member, and make him wonder if you are covering up a weakness by a false show of strength.

GET COMFORTABLE: don't lounge or sprawl, sit erectly but not stiffly. A careless posture may lead the board to conclude you are careless in other things, or at least that you are not impressed by the importance of the interview. Don't fuss with your clothing or with a pencil or ash tray, etc. Your hands may occasionally be useful to emphasize a point; don't let them detract from your presentation by becoming a point of distraction.

DON'T WISECRACK: or make small talk, this is a serious situation, and your attitude should show that you consider it as such. Further, the time of the board is limited; they don't want to waste it, and neither should you.

DON'T EXAGGERATE: your experience or abilities, the board has your application in front of them and also from other sources may know more about you than you think. Also you probably won't get away with it anyway. An experienced board is rather adept at spotting such a situation. Don't take a chance!

IF YOU KNOW A MEMBER OF THE BOARD: don't make a point of it, but don't hide it. Certainly you are not fooling him, or the other members of the board. Don't try to take advantage of your acquaintance with this individual, it will likely hinder your score.

DON'T ATTEMPT TO DOMINATE THE INTERVIEW: Let the board have control. They will give you clues, don't assume that you have to do all of the talking. Be aware that the board has a number of questions to ask you. Don't try to take up all the interview time by showing off your extensive knowledge of the answer to the first question that you are asked.

BE ATTENTIVE: you will have a limited time in the interview, keep your attention at a sharp level throughout this period. When a board member is questioning you, give this person your undivided attention. Direct your response primarily to the board member that ask the question, but don't exclude the other board members.

DON'T INTERRUPT: A board member may be stating a problem for you to analyze. When the time comes you will be asked a question. Let the board member state the problem, and wait for the question.

MAKE SURE THAT YOU UNDERSTAND THE QUESTION: Don't try to answer until you are sure what the question is. If the question is not clear, restate it in your own words or ask the board member to clarify it for you. Don't haggle about minor elements.

REPLY PROMPTLY: But don't reply hastily. A common entry on oral board rating sheets is " candidate responded readily" or "candidate hesitated in replies". Respond as promptly and quickly as you can, but don't jump to hasty, ill-considered responses.

DON'T BE PREEMPTORY IN YOUR ANSWERS: A brief answer is proper, but don't fire your answers back. This is a losing game from your point of view. The board member can probably ask questions much faster than you can answer them.

DON'T TRY TO CREATE ANSWERS THAT YOU ASSUME THAT THE BOARD WANTS TO HEAR: The board is interested in what kind of a mind you have and how it works, not in playing games. Most board members can spot this tactic and will grade you down for it.

DON'T CHANGE VIEWS IN ORDER TO PLEASE THE BOARD: Board members will take a contrary position in order to draw you out and see if you are willing and able to defend your point of view. Don't start a debate, but don't surrender a good position. If your position is worth taking, it is worth defending.

IF YOU ARE SHOWN TO HAVE MADE AN ERROR IN JUDGEMENT, DON'T BE AFRAID TO ADMIT THE ERROR: The board that you are forced to reply without any opportunity for careful consideration. Your answer may be demonstrably wrong. If so, admit it and get on with the interview.

DON'T SPEND TOO MUCH TIME DISCUSSING YOUR PRESENT JOB: The opening statements may concern your present employment. Answer the question but don't go into an extended discussion. You are being examined for a new job, not your present one. As a mater of fact, try to phrase all you answers in terms relating to the position of Firefighter.

DON'T TELL STORIES: Keep your responses to the point. If you feel the need for illustration from your personal experience, keep it short. Leave out the minor details. Make sure that the incident is true.

DON'T BE TECHNICAL OR BORING: The board is not interested in ponderous technical data at this time.

DON'T USE SLANG TERMS: many a good response has been weakened by the injection of slang terms or other jargon. Oral boards usually will notice any slips of the grammar or any other evidence of carelessness in speech habits.

DON'T BRING DISPLAYS OR DEMONSTRATIONS: The board members are not interested in letters of reference etc.

DON'T BE INGRATIATING: This routine rarely works with an oral board. Be pleasant and smile occasionally, but do it naturally and don't overdo it.

SOME WAYS THAT CAN CAUSE YOU TO "STRIKE OUT" IN AN ORAL INTERVIEW:

1. Poor personal appearance.
2. Lack of interest and enthusiasm.
3. Passiveness or indifference.
4. Overemphasis on wages.
5. Condemnation of past employers.
6. Failure to look at board members during interview.
7. Limp, fishy handshake.
8. Indefinite response to questions.
9. Overbearing, overaggressive, conceited with superiority or "know it all" attitude.
10. Inability to express self clearly: poor voice, diction, or grammar.
11. Lack of planning for position.
12. Lack of confidence and poise: nervous, ill at ease.
13. Make excuses: evasive; hedges on unfavorable factors in work record, etc.
14. Lack of tact.
15. Lack of courtesy; ill mannered.
16. Lack of maturity and/or vitality.
17. Indecision.
17. Sloppy application.
19. Merely "shopping" for position.
20. Wants position only for a short time.
21. Lack of interest in jurisdiction, appear lazy.
22. Low moral standards.
23. Intolerant: strong prejudices.
24. Narrow interest.
25. Inability to take criticism.
26. High pressure type.
27. Inability to listen, domination of the interview.

AFTER THE ORAL INTERVIEW

After you are excused form the ORAL INTERVIEW, go somewhere that you feel comfortable and make a list of:

1. How you felt during the interview.

2. Any question/questions that you feel you may want to recall at a later date.

3. Types of questions asked.

4. How questions were asked.

5. How the oral board reacted to your answers.

6. How you may have improved your preparation for this exam.

XXXXXXXXXXXXXXXXXXXXXXXXXXXXXXXXXXXXXXXXXXXXXXXXX

FOR INFORMATION CONCERNING FIRE SERVICE ORAL INTERVIEWS, ASSESSMENT CENTERS AND FIRE SIMULATOR EXAMS, SEE SECTION #3.

For a more complete explanation of FIRE SERVICE ORAL INTERVIEWS, ASSESSMENT CENTERS and SIMULATOR EXAMS, along with a complete description of each exercise see: **"FIRE CAPTAIN ORAL EXAM STUDY GUIDE"**.
Available from **INFORMATION GUIDES**.

XXXXXXXXXXXXXXXXXXXXXXXXXXXXXXXXXXXXXXXXXXXXXXXXX

SECTION #6

FIRE AND FIREFIGHTING PRINCIPLES

FIRE SCIENCE

Every thing is made up of atoms of the basic elements. Atoms combine with the other elements to form molecules. This is matter and matter cannot be destroyed but merely changed into other matter or into energy.

Three forms of matter:

1. Solid, packed molecules, the molecules are close together along with great common attraction.

2. Liquid, loose molecules, the molecules are farther apart with less common attraction.

3. Gas, very loose molecules, the molecules are very far apart with the least amount of common attraction.

ATOM: the smallest particle of an element that can exist alone or in combination.

HEAT: energy that is associated with and proportional to molecular motion, and that can be transferred from one body to another by radiation, conduction, and convection.

FIRE: rapid self-sustaining oxidation accompanied by the evolution of varying intensities of heat and light.

Heat is measured in British Thermal Units. This is not the same thing as temperature. Temperature is the heat at the moment. BTU tells how much heat there is in a substance. A large substance will most likely have a greater heating capacity than a small one, even if they have the same temperature.

Heat always flows from the higher temperature substance to the lower temperature substance. This gives Firefighters three ways to the heat of a fire off a substance, by:

CONDUCTION, the heat will flow off, which causes fires and can also be used to extinguish fires. When the heat from a hot substance goes into a flammable substance, that it is in contact with, a fire may follow. If the heat of the burning substance is transferred into water, it will be cooled, which may extinguish the fire.

80

CONVECTION, the heat will travel in the air or water. A fire in lower floors of structures will travel up stairways by convection to higher areas within the structure. When proper ventilation techniques are used, the heated gases will be expelled to the outside of the structure.

RADIATION, the heat travels at the speed of light just as any other beam. If the heat is stopped by an opaque substance, it would then be transferred by conduction.

FIRE SCIENCE the knowledge concerning the behavior, effects, and control of fire.

FIRE TRIANGLE: a three sided figure representing the three of the four factors necessary for combustion.

1. Oxygen.
2. Heat.
3. Fuel.

FIRE TETRAHEDRON: the four elements required by a fire:

1. Fuel.
2. Heat.
3. Oxygen.
4. Uninhibited chain reaction.

Three techniques for extinguishing fire:

1. Remove the fuel.
2. Reduce the temperature. (water is the best extinguishing agent)
3. Eliminate the oxygen.

Reactions developing during COMBUSTION and BURNING may be characterized as CHAIN REACTIONS.

FLAMES: visible flickering light, various colors.

FLAME ENVELOPE: the boundary of process that transforms the fuel and air mixture into fire.

FLAME: the burning gases or vapors from fire.

SPARK: a tiny, bright speck from burning material, or radiant fragments created from grinding metal.

FIRE CAUSE: is a condition that starts a fire or allows a fire to start.(the source of ignition).

FLASHOVER: a fire continues to burn all the contents gradually heating to their ignition temperatures.

FIRE SPREAD: the involvement and migration of fire across surfaces. (flame spread)

FLASHOVER POINT is when a fire consisting of the interior structure and contents, spreading slowly, with ample combustibles and air present so as to generate the required temperatures that will eventually reach the point where all combustible surfaces will burst into flame.

FLAMEOVER: the rapid spread of fire over one or more surfaces.

FIRE STORM: a violent, convective atmospheric disturbance caused by large intense fires which tend to suck all the available air into the fire.

INCENDIARY FIRE: is a fire that is set wilfully.

The chemical stability of MOLECULES determines the manner in which oxygen will unite with COMBUSTIBLE materials.

Smoke may reach temperatures of over
1000 DEGREES F.

FIRE WIND: the wind generated by an intense fire that consumes oxygen from the atmosphere which develops a incomplete vacuum and generates the motion of additional air in the direction of the fire.

AIR and CARBON MONOXIDE mixtures will not ignite at temperatures of less than 1100 DEGREES F.

FIRE GAS: a gaseous product of combustion.

GASES: a quantity of matter existing in a gaseous state at ordinary temperature and pressure.

GASES expand indefinitely.

As a fire elevates in temperature, the proportion of CARBON MONOXIDE in the gases released enlarges.

ADIABATIC: change in the condition of a substance that will take place without the gain or loss of heat from surrounding materials.

CALORIE = one gram of water raised 1 degree C.

CALORIES and BTU'S are measurements of heat.

AMBIENT TEMPERATURE relates to temperatures encompassing on all sides.

Liquid molecules are consistently in motion, the amount of motion is mostly subject to the liquid's TEMPERATURE.

The principle of the EXPANSION OF LIQUIDS is used in the THERMOMETER: which uses a glass tube with Mercury or Alcohol which rises or falls as they expand or contract from heat change.

1 degree C = 1.8 degrees F - (+32 degrees F).
Example: 50C = 90F + 32F = 122 degrees F.
THIS IS FOR POSITIVE TEMPERATURES!

1 degree C = 1.8 degrees F - (+32 degrees). Example: -40C = 72F -(+32F) = 40 degrees F.
THIS IS FOR NEGATIVE TEMPERATURES!

-40 degrees F and -40 degrees C is the only temperature that degrees F and degrees C are IDENTICAL.

HIGH HUMIDITY tends to keep the products of combustion from rising into the atmosphere.

RELATIVE HUMIDITY: the ratio of the actual moisture content of air to its saturation content at that temperature.

WATER : a compound consisting of one part Oxygen and two parts Hydrogen.

Temperature is measured by thermometers. There are two scales that are used: Fahrenheit scale and Centigrade scale. The Fahrenheit scale registers freezing point as 32 degrees, and boiling point of water at 212 degrees making a difference of 180 degrees between the two points. Centigrade uses 0 degrees as the freezing point and 100 degrees as the boiling point of water. To convert one to the other remember that Fahrenheit is 9/5 Centigrade plus 32 and/or Centigrade is 5/9 of Fahrenheit. (Fahrenheit minus 32).

MISCIBILITY: the ability of a flammable liquid to mix with water.

A HOMOGENOUS-FUEL-OXYGEN mixture is uniformly mixed.

RAPID OXIDATION: Oxygen and a burning material along with the aid of heat.

COMBUSTION: rapid oxidation.

PH SCALE:

1. Above 7.0 = bases : usually solid.(alkalis)

2. Below 7.0 = acid : usually liquid.

3. At   7.0 = Neutral.

SPECIFIC GRAVITY = the weight or mass of a given volume of a substance at a specified temperature, as compared to that of an equal volume of another substance.

Liquids of a specific gravity of less than 1 will FLOAT ON WATER.

SPECIFIC GRAVITY applies to liquids only.

VAPOR - GASES:

FLAMMABLE DENSITY: is the range of combustible vapor or gas mixtures with air between the upper and lower flammable limits.

Flammable liquid vapors have VAPOR DENSITY of greater than 1, air = 1 therefore they will sink in the atmosphere.

VAPOR DENSITY = the weight of a vapor-air mixture resulting from the vaporization of a flammable liquid at equilibrium temperature and pressure conditions, as compared with the weight of an equal volume of air under the same conditions.

In order to compare VAPOR DENSITIES of gases, you must know that the MOLECULAR WEIGHT OF AIR = 29.

The VAPOR PRESSURES of flammable liquids are usually expressed in POUNDS PER SQUARE INCH. (PSI)

VOLATILE: capable of being evaporated/vaporized easily.

VAPORIZATION: is the process that a substance changes from liquid or solid phase to a gas.

VAPOR PRESSURE: is the point that a liquid within a closed container equalizes the rate of the molecular escape with the rate of molecular return to the fluid.

The RATE OF DIFFUSION of an unconfined gas varies inversely with its VAPOR DENSITY.

VAPOR PRESSURE OF A LIQUID: is the pressure of the vapor at any given temperature at which the vapor and liquid phases of the substance are in balance in a closed vessel.

FLAMMABLE GASES and VAPORS will only burn or explode if the proportion of gas or vapor is within certain maximum or minimum limits.

Before any solid flammable substance can burn, it must be in the form of VAPOR.

BOYLE'S LAW: if the temperature remains constant, the volume of gas varies inversely with pressure.

The physical properties of VAPORS and GASES are closely related.

SUBLIME: is when a solid changes to vapor without passing through the liquid phase.

STEAM: the visible water vapors forming from the boiling of water

When converting liquid to vapor, the VOLUME IS INCREASED BY 1700 TIMES.

A gallon of water may produce a maximum of 200 CUBIC FEET OF STEAM.

LIQUIDS will not expand indefinitely like gases.

A SUBSTANCE = gas if it is in a gaseous state at 100 degrees F and 40 PSI.

A SUBSTANCE = liquid if it is in a liquid state at 70 degrees F and 14.7 PSI.

IGNITION TEMPERATURE:

IGNITION TEMPERATURE = temperature that a SUBSTANCE is able to continue to burn without outside heat.

IGNITION TEMPERATURE is the temperature at which a SOLID COMBUSTIBLE SUBSTANCE will ignite and burn.

Factors affecting the IGNITION TEMPERATURE of a combustible solid are:

1. Size of the solid.
2. Shape of the solid.
3. Rate of heating the solid.

A solid combustible substance will ignite and burn when it has been heated to its IGNITION TEMPERATURE.

FLASH POINT-FIRE POINT:

A liquid is classified flammable depending upon its FLASH POINT.

The lowest temperature that a flammable liquid provides adequate vapor to form an ignitable mixture with air close to the surface or within a vessel is the FLASH POINT.

150 DEGREES F is the lowest FLASH POINT temperature of a flammable liquid that it is practical to use a water spray nozzle to extinguish a fire.

FLASH POINT more than any other property determines the hazard of the liquid.

FLASH POINT = less than 5 DEGREES BELOW fire point.

Most USEFUL information concerning hazard of a liquid is the FLASH POINT.

FIRE POINT is the lowest temperature of a LIQUID at which vapors are evolved fast enough to continue combustion.

HEAT/HEAT TRANSMISSION:

HEAT: the energy affiliated with and balanced to the molecular motion which can be transferred from one substance to another by:

1. Radiation.

2. Conduction.

3. Convection.

HEAT OF COMBUSTION: measure of HEAT liberated while a substance is going through complete oxidation.

HEAT OF VAPORIZATION: amount of HEAT needed to convert a unit quantity of liquid into vapor.

HEAT CAPACITY: the quantity of HEAT needed to raise a given unit of substance one degree without phase of chemical changes. (Specific Heat)

The SPECIFIC HEAT of a material describes the materials ability to absorb heat.

SPECIFIC HEAT = number of B.T.U.'s to raise 1 LB substance 1 degree F.

SPECIFIC HEAT = absorption of heat.

The SPECIFIC HEAT of water is always higher than other common substances.

LATENT HEAT = heat absorbed or given off from a substance as it transfers from a liquid to a gas or a solid to a liquid.

EXOTHERMIC HEAT: gives up heat.

ENDOTHERMIC HEAT: absorbs heat; chemical reaction.

ENDOTHERMIC REACTION: a process or change that absorbs heat and requires it for initiation and maintenance.

EXOTHERMIC: designating a chemical change in which heat is released.

HEAT is released from a substance when it converts form a gas form to a liquid form.

CHEMICAL HEAT REACTIONS:

1. Heat of decomposition.

2. Heat of solution.

3. Spontaneous heating.

The first five minutes of a fire are the most important because HEAT accelerates chemical reaction.

B.T.U. = British Thermal Unit.

B.T.U. is the amount of heat required to raise one pound of water one degree F. (at atmospheric pressure).

B.T.U. = 1 LB water increases 1 degree F.

To convert one pound of ice at 32 degrees F to steam at 212 degrees F requires 1293.7 B.T.U.'s.

To convert ice to water requires 143.4 B.T.U.'s.

To convert water to steam requires 970.3 B.T.U.'s.

To raise 32 degrees F to 212 degrees F requires 180 B.T.U.'s.

One gallon of water will absorb about 8000 B.T.U.'s.

BOILING POINT = temperature when vapor is equal to atmospheric pressure.

BOILING POINT = the temperature that a liquid will swiftly convert to vapor.

Generally as the BOILING POINT of a liquid goes down, the VAPOR PRESSURE and evaporation rate INCREASE.

BOIL: when the temperature of a fluid is raised to a point that the vapor pressure is equal to the atmospheric pressure.

Water heated at sea level will BOIL at a higher temperature than water heated at elevated areas.

BOILOVER in flammable liquid storage tank fires is most directly related to the water content of the liquid.

BOILOVER of a burning oil tank is an indication that water is present either in suspended form or at the bottom of the tank.

BURNING RATE: the speed that solids or liquids will burn.

The quantity of HEAT TRANSFER through a substance:

1. Directly proportional to cross-section area.

2. Directly proportional to temperature difference between two points in the substance.

3. Inversely proportional to the distance the heat travels.

BURNING CONDITIONS: the blend of surrounding circumstances that will influence any fire in any given fire situation:

1. CONDUCTION; direct heat contact.

2. RADIATION; in all directions where matter does not exist, such as air.

3. CONVECTION; by air currents usually in an upward direction.

4. DIRECT flame contact.

SPONTANEOUS IGNITION-OXIDATION:

SPONTANEOUS HEATING and IGNITION develop when the inherent properties of a material cause exothermic chemical reactions to progress without exposure to an external heat source.

Substances that are susceptible to SPONTANEOUS COMBUSTION are able to catch fire without an external source of heat.

In all cases of SPONTANEOUS IGNITION heat of oxidation must be produced more rapidly than it is dispersed.

SPONTANEOUS = slow oxidation.

SPONTANEOUS HEATING is heating due to chemical or bacterial action in a combustible material.

SPONTANEOUS IGNITION is ignition due to chemical reaction or bacterial action in which there is a slow oxidation of organic compounds until the material ignites; usually there is sufficient air for oxidation but not enough ventilation to carry heat away as it is generated.

OXIDATION a chemical reaction in which Oxygen combines with other substances.

OXIDATION any chemical reaction in which electrons are transferred.(always produces heat)

In FIRE CHEMISTRY, combustible fuels which contain CARBON and HYDROGEN are called REDUCING AGENTS.

OXIDATION and reduction always occur simultaneously, and the substance which gains the electrons is called the OXIDIZING AGENT.

MAGNESIUM and SODIUM are metals that are classified as REDUCING AGENTS.

FIRE STAGES - BACKDRAFT - COMBUSTION:

FIRE BURNS IN TWO MODES:

1. Flaming mode   = Tetrahedron the second phase is free burning.

2. Smoldering mode = Fire triangle the third phase is smoldering.

STAGES OF FIRE:

1st = oxygen at 21%, fire at 1000 degrees F, room at 100 degrees F.

2nd = oxygen at 21% to 15%, fire and room at 1300 degrees F.

3rd = oxygen below 15%, fire and room at 1000 degrees F.

FIRST STAGE OF FIRE = smoldering, incipient phase.

SECOND STAGE OF FIRE = flame producing phase.

THIRD STAGE OF FIRE = smoldering phase.

SMOLDERING PHASE of fire = decrease in heat generation.

During the FIRST STAGE OF FIRE:

1. Little or no decrease in the oxygen content of the interior atmosphere.

2. Little or no increase in average temperature of the interior atmosphere.

3. Major damage will be caused by smoke.

During the SECOND PHASE OF FIRE is when the major destruction will take place.

During the THIRD PHASE OF FIRE there is a possibility of an inward rupture of window panes.

If a Firefighter opens the door to an apartment and a smoke explosion occurs, the fire was most likely in the THIRD PHASE OF FIRE.

The SMOLDERING PHASE of a confined fire within a building is characterized by the decrease in heat generation.

Fires usually burn UPWARD AND OUTWARD.

The primary action of NITROGEN in the atmosphere, during a fire, is to slow up the burning reactions.

With each 18 DEGREES F rise in temperature, chemical reactions will double their speed or rate.

ATMOSPHERIC AIR CONTAINS:

1. 21% Oxygen.
2. 78% Nitrogen.
3. 1% Miscellaneous gases.

BACKDRAFT CHARACTERISTICS:

1. Smoke under pressure.
2. Dense grayish, yellowish smoke.
3. Puffing smoke from cracks, moving up rapidly.
4. Confinement of excessive heat.
5. Sweating windows, hot to the touch, and dark in color.
6. Muffled sounds.
7. No visible flame.
8. Rapid movement of air inward when opening is made.

One of the BEST indications that a confined fire may be causing a potential BACKDRAFT CONDITION is thick black smoke, sometimes grayish and yellowish.

A LONG SMOLDERING FIRE may be indicated by:
1. Smoke stains on window glass.
2. Cracks in window glass, with varnish like or flat paint appearance.

Common combustible materials will not burn if the OXYGEN PERCENTAGE drops to 15% or below.

Smoke normally RISES up from fire because cooler, heavier air displaces the lighter warmer air.

An exception to the accepted rule that SMOKE RISES is a fire in the upper part of a cold storage plant where the smoke comes into contact with the refrigerating system.

MUSHROOMING: the condition when the heat and gases spread out laterally at the top of a structure.

MUSHROOMING of a fire is most likely to occur in any building where the heat is confined to the upper floors.

PRODUCTS OF COMBUSTION:
1. Fire gases; Oxygen, Hydrogen, and Carbon.
2. Flame, heat, and smoke.

When a firefighter enters a building involved in fire, he can expect to find the PRODUCTS OF COMBUSTION, depending on what is burning.

COMBUSTION: a rapid exothermic oxidation process accompanied by continuous evolution of heat and usually light.

COMBUSTION usually occurs after vaporization and decomposition due to heat.

FIREFIGHTING

SIZE-UP = reconnaissance.

SIZE-UP : The mental evaluation made by the fire officer in charge, which enables him to determine a course of action; it includes such factors as time, location, nature of occurrence, life hazard, exposure, property involved, nature and extent of the fire, available water supply and other fire fighting facilities.

SIZE-UP is a report usually via radio, giving existing conditions of an emergency.

SIZE-UP: the survey of the situation that is taken upon arrival at the scene of an incident.

SUCCESS or FAILURE on the fire ground depends a lot on the ability of a fire officer to SIZE-UP the situation skillfully and quickly survey situation changes that develop.

FOUR STAGES of SIZE-UP:

1. Anticipating the situation.
2. Gathering the facts.
3. Evaluating the facts.
4. Determining the procedures.

Things to consider at the time of SIZE-UP:

1. Location of fire.
2. Location of fire within a structure.
3. Size of fire.
4. Smoke and gases being generated.
5. Type of contents within a structure. (rags, flammable oils, plastics, etc.)
6. Potential or actual danger to firefighters.
7. Potential or actual danger to occupants of a structure.

MOST IMPORTANT consideration of SIZE-UP is location.

SIZE-UP is continuous throughout the entire fire fighting operation.

FIRE TACTICS: various maneuvers that can be used in a strategy to successfully fight a fire.

FIRE STRATEGY: the plan of attack on a fire.

FIRE STRATEGY: should make prime use of equipment and personnel, and take into consideration fire behavior, the nature of the occupancy, environmental conditions, and weather factors.

Even though NO TWO FIRES ARE ALIKE, it is possible to lay down general plans for firefighting operations primarily because the elements of similarity are sufficient enough to establish TACTICS and STRATEGY applicable in a variety of situations.

For covering all points of a fire remember : "FRONT and REAR, OVER and UNDER, and COVER ALL EXPOSURES".

FOUR MAJOR OBJECTIVES of Fire Departments:

1. Prevent loss of life and property.
2. Extinguish fires.
3. Prevent fires.
4. Confine fires to point of origin.

THEORY can be best aid for lack of experience.

The most common mistake that first in officers make is that they do not request ADDITIONAL HELP.

When the first on scene officer becomes aware that a building fires potential may go beyond the capabilities of the available apparatus and manpower, his first action should be to CALL FOR MORE HELP.

SOME FACTORS THAT MAY BE USED IN DETERMINING THE NEED FOR MORE MANPOWER:

1. Possibility of the fire spreading.
2. The number of hose lines needed.
3. The type and nearness of exposures encircling the fire area.

The first in fire officer should calculate WHERE THE FIRE IS TRAVELING, immediately after he has discovered the seat of the fire.

When trying to establish which WAY THE FIRE IS TRAVELING, the first and most important consideration is the interior stairway so that the path of occupants will not be hindered.

The most practical technique of managing and extinguishing fires involving ORDINARY COMBUSTIBLES in structures is to transfer the excessive heat from the involved and exposed combustibles to a non-combustible substance, such as water, etc.

In the Fire Service, the key to the TASK FORCE concept is COMMAND.

SPEED OF DISCOVERY = most important to fire loss.

MANPOWER is most critical at early stages of fire.

In any firefighting, the FIRST OPERATION to consider is the LIFE HAZARDS.

The PRIMARY FUNCTION of an engine company is to OBTAIN and DELIVER WATER.

PRIMARY DUTY of an officer of a pump company, AFTER assignment to tactical position, is "SETTING UP THE PUMP" at a suitable hydrant or water source.

The BASIC Fire Department unit is the FIRE FIGHTING COMPANY.

The BASIC JOB of a pumper company is to get water on fires quickly, efficiently, in the right places, in the right amounts.

Some BASIC ROLES of a Firefighting company:

1. Fire attack.
2. To supply sufficient volume of water at the correct pressures for the necessary fire streams.

PRE-FIRE-PLANNING is the specific responsibility of fire suppression officers.

One piece of apparatus cannot accomplish two different basic FIRE TACTIC functions at the same time.

Unsatisfactory results in firefighting will take place when fire companies are used as INDIVIDUAL UNITS at the fire ground.

COMPONENTS OF A BASIC FIRE DEPARTMENT RESPONSE:

1. Appropriate manpower.
2. Attack hose lines of various sizes.
3. Heavy stream capabilities.
4. Water pumping capabilities.
5. Ladder capabilities.
6. Forcible entry capabilities.
7. Ventilation capabilities.
8. Exposure coverage capabilities.

The objective of FIREFIGHTING in places of PUBLIC ASSEMBLY include:

1. Prompt evacuation of persons within structure.
2. Covering of exits with fire streams for safe exit of persons within structure.
3. Ventilation of structure or fire area.
4. Maintain unobstructed egress facilities.
5. Maintain extinguishment of fire.

FIREFIGHTING SEQUENCES-METHODS:

SEQUENCE OF FIRE FIGHTING:

1. Locate the fire.
2. Confine the fire.
3. Extinguish the fire.

TO FIND FIRES IN WALLS:

1. Look. 2. Listen. 3. Feel.

The strategy of THREE - PRONGED or THREE POSITION ATTACK is necessary to control and extinguish any fire of any magnitude.

The three positions of THREE - PRONGED or THREE POSITION ATTACK are:

1. Exposures.
2. Avenues of fire spread.
3. The seat of the fire.

If another hydrant is available, pumpers should NEVER be connected to a hydrant directly in FRONT of burning structure.

It is the function of PUMPING APPARATUS to produce the appropriate pressure needed for specific hose lines, from the water obtained from fire HYDRANTS.

It is the function of the HYDRANT SYSTEM to supply the VOLUME of water needed at a fire scene.

At a large fire, it is preferable for a second in fire pumper to hook up to the SAME HYDRANT as the first in apparatus near the fire, rather than a block away, because the main will carry water with much less friction loss than supply hose lines.

When laying lines at a fire, the lines should be commensurate with the SIZE OF THE FIRE.

LARGE VOLUMES OF WATER are required to control and extinguish an open or unconfined building fire.

SPOTTING fire apparatus close to the fire structure, at large fires, is proper since access to hose loads and equipment will be more efficient.

Small fire streams = MANEUVERABILITY.

The PRIMARY use of a BOOSTER TANK on a pumper is that it makes it possible to control the majority of small fires with minimum water damage.

A basic rule on any structure fire is CONFINEMENT.

POOR hose lays = low capacity.

AMOUNT OF WATER for a fire depends on the amount of heat generated.

Hose lines usually should not be charged until they are at the point WHERE THE STREAM IS TO BE USED.

DURING FIREFIGHTING OPERATIONS REMEMBER:
1. With the same nozzle pressure the vertical range of a stream will be greater than the horizontal range.
2. An increase in nozzle pressure will be accompanied by an increase in nozzle reaction.
3. With the same nozzle pressure in each case, a larger tip will have greater range than a smaller tip.
4. An increase in nozzle pressure will be accompanied by an increase in flow.
5. A nozzle that is too large for a specific hose diameter will give a weak and ineffective fire stream.
6. Any wind regardless of direction will hinder a fire stream.

LARGEST NUMBER of fire streams that first in engine company can STRETCH and OPERATE is TWO.

Hose lines of over 400 - 500 FEET are inefficient and slow down operations.

When advancing hose lines for a fire on the third floor of a structure, pull enough to reach 200 FEET past the front entrance.

The LOCATION OF THE FIRE HYDRANT is the determining factor as to whether or not to lay hose lines forward or reverse.

It is not recommended to advance upon an fire INSIDE A STRUCTURE from different directions, because it will be difficult to get close to the seat of the fire with the smoke and heat that will be driven back and forth.

If heat, smoke, and flames make it unsuitable for Firefighters to advance hose lines on the leeward side of a large fire, the attack should be made with heavy streams on the FLANK.

Atmosphere EXCEEDING 120 - 130 DEGREES F, use protective clothing.

When fighting an EXTERIOR FIRE, the FIRE OFFICER should first evaluate the situation, and then be aware as how to control the spread of the fire.

When a fire is under control, QUESTIONABLE AREAS should be observed carefully, to determine whether any other action is needed.

At all large fires, HEAVY STREAM equipment should be in place and ready for service in the event that, the incident commander thinks that the hand lines will not be able to control the situation.

HEAVY STREAM placement should include coverage of exposures as well as cooling of the fire, if possible.

HEAVY STREAM equipment should be set up before hand lines prove inadequate.

If two or more pumpers must be used in RELAY OPERATIONS, the pumper with the largest capacity should be the one closest to the water source.

RELAY PUMPING is the use of two or more pumpers so as to pump water over a extensive distance by operating the pumpers in series.

INDIRECT FIRE ATTACK:

INDIRECT FIRE ATTACK is the technique of firefighting, that was originated by FIRE CHIEF LLOYD LAYMAN, of introducing water particles into a heated atmosphere over a fire as opposed to applying water directly on the fire.

During INDIRECT FIREFIGHTING the water particles convert into steam and expand in volume by about 1 to 1800, and they cool and smother the fire simultaneously.

The principle of fire extinguishment with the use of WATER SPRAY or "FOG" is based primarily on the point that water consumes large amounts of heat when it evaporates.

Theoretically, one gallon of water at 90% efficiency will form enough steam to cool a fire zone of 100 CUBIC FEET within 30 SECONDS.

The principle of an INDIRECT ATTACK, is the transferral of heat from combustible materials to non-combustible substances.

An INDIRECT FIRE ATTACK will move the excessive heat from the inside of a structure to the outside atmosphere.

When attacking a fire with the INDIRECT method, the initial attack should be made in the area of the greatest involvement.

When making an INDIRECT ATTACK on a fire, it is more sensible to make the attack from the outside, if you appraise the hazards and effectiveness.

The effectiveness of an INDIRECT ATTACK depends primarily upon continuing the flow of water particles, without stopping, until the amount of steam is notably reduced.

The principal reason for converting water to steam in an INDIRECT FIRE ATTACK is to acquire the maximum cooling effect of water.

SPRAY will protect an exposure from radiated heat.

SPRAY will carry water toward a fire that is inaccessible.

SPRAY STREAMS cooling effect is more effective than straight streams because they present a greater surface area.

The full cooling capacity of water is being employed efficiently at a fire when all of the water is transferred into STEAM.

EXPOSURES:

EXPOSURE is the situation when a property is put in danger by a fire involved in another structure or outside nearby.

EXPOSURE FIRE is a fire that is caused by another fire.

EXPOSURE HAZARD is the speculation that a structure or area will catch fire or be damaged by a fire from an adjoining property.

Aside from life safety, the primary task of first in units is to supply appropriate protection for the EXPOSURES.

When no lives are in immediate danger, the first fire streams should be placed to thwart any EXPOSURES from becoming involved.

COVERING EXPOSURES: the operations that are necessary to prevent the extension of fire from an involved structure to other structures or areas.

The best method to protect EXPOSURES is to extinguish the fire before it gets away.

FUNDAMENTAL RULES IN PROTECTING EXPOSURES:

1. Send men, not lines.
2. Control the fire and the exposures will take care of themselves.
3. Control the exposures and the fire will take care of itself.

In general it is a mistake to only keep water on the EXPOSURES when nothing is done to cool the main body of a fire structure, because the heat produced could render the DEFENSE INEFFECTIVE.

EXPOSURE PROTECTION is the employment of water spray to a structure or an area that is not burning but is exposed to fire, so as to cool or limit the heat absorption and prevent ignition.

Officer in charge of fire fighting operations at a building fire should plan to first cover EXPOSURES on the LEEWARD SIDE of the fire structure.

The protection of EXPOSED STRUCTURES includes:
1. Covering the leeward side first.
2. Placing fire streams between involved structure and exposed structures.
3. Using fire streams designated for exposed structures on fire structure.

The most dangerous side of a fire structure is the LEEWARD side.

The lack of EXPOSURE protection may cause a CONFLAGRATION.

CONFLAGRATION: a large fire that envelops a substantial area and is able to cross natural obstructions.

Consider WATER DAMAGE when fighting fires:
1. On the upper floors of a structure.
2. In buildings containing delicate instruments.
3. On the first floor of a building with a basement.
4. In a sugar warehouse. Etc.

FIRE EXTENSION is the spread of fire to other areas that were not previously involved.

FIRE EXTENSION:

FIRE EXTENSION: the spread of fire to different areas not already involved in fire.

FIRE ENDURANCE: duration of time a substance or product element, structure, etc., will sustain its integrity during a fire.

Fires are most frequently spread by CONVECTION.

PROPAGATION OF FLAME is the spread of flame from layer to layer independently of the source of ignition.

HORIZONTAL EXTENSION OF FIRE MAY OCCUR:

1. Through wall openings by direct flame contact.

2. Through open space by radiated heat or by convected air currents.

3. Through walls and interior partitions by direct flame contact.

DOWNWARD EXTENSION OF FIRE MAY OCCUR:

1. Through floors by direct flame contact.

2. Through ceilings by direct flame contact.

Fire, heat, and smoke will only travel downward when the upward and horizontal routes are BLOCKED.

DAYS LEAST FAVORABLE TO FIRES:

1. Relative humidity less than 40%

2. Precipitation less than .01 the day of fire

3. Precipitation less than .01 three days prior to fire.

4. Maximum wind speed is no more than 13 MPH.

WEATHER: the general state of the atmosphere at a specific time and place, with respect to temperature, cloudiness, moisture, etc.

FIRE DEVIL: a small, burning cyclone that occurs usually during forest and brush fires but can also occur in free burning structural fires. Fire devil will form a vortex as heated gases from a fire rise and cooler air rushes into the resulting areas of low pressure.

CONFLAGRATION HAZARD: is a close group of structures that are subject to rapid fire spread.

FUEL greatly influences fire behavior.

Burning metal temperatures are MUCH HIGHER than burning flammable liquids.

Smoke from plastic's involved in fire, produce at least TWICE AS MUCH as in a wood fire.

PLASTIC DUST is the easiest form of plastic to ignite.

Flammability of PLASTICS varies with the shape and form of the plastic.

The greatest FIRE DAMAGE for plastic materials is in the FOAMED STATE.

Most common spread of fire in buildings is UNPROTECTED VERTICAL OPENINGS.

Firefighting within a structure from several different directions can be very difficult mainly because the SMOKE and HEAT will be driven from one direction to another.

The most significant factor for spreading fire in structures are STAIRS and SHAFTS.

In most structure fires, the floor temperatures are usually about ONE THIRD that of the ceilings.

TRUCK COMPANY OPERATIONS:

1/3 of the manpower at an average fire usually involves TRUCK and RESCUE OPERATIONS.

The primary FUNCTION of a LADDER COMPANY is RESCUE.

Primary DUTIES of a LADDER COMPANY:

1. Search and rescue.
2. Ventilation.

TRUCK COMPANY DUTIES:

1. Rescue.
2. Forcible entry.
3. Ventilation.
4. Ladder evolutions.
5. Use of various tools and equipment.

RESCUE procedures at a fire of a multistory structure, on upper floors, can usually best be carried out by the use of STAIRWAYS.

VENTILATION: a planned and systematic release and removal of heated air, smoke, and gases from a structure and the replacement of these products of combustion with a supply of cooler air.

PRINCIPLE of VENTILATION is to let heat and smoke out and bring in cooler air so that Firefighters can work inside a structure.

There are no FIXED rules for opening or VENTILATING a fire.

VENTILATION at a fire:

1. Thwarts additional ignition of contents by withdrawing heat and gases.
2. Tends to move the fire in a course where its spread may be managed.
3. Can greatly decrease the amount of heat that must be cooled by hose streams.

At all structure fires VENTILATION should not start until the hose lines are in position, manned and charged with water.

Roof VENTILATION of a tightly confined structure fire should not begin until there are charged hose lines in place at the entrance ready for action.

VENTILATION may be required at any time following the initial size-up.

Opening the roof over a vertical shaft will usually cause an UP DRAFT action of ventilated gases.

Conventional methods of VENTILATION should take place during the first and/or second phase of the fire modes. NOT during the third phase.

DIRECT DISPLACEMENT: conventional method of VENTILATION.

SAFETY OF THE OCCUPANTS is the first consideration of ventilation.

When a structure is well involved with smoke, try to VENTILATE at the highest point, avoid VENTILATING from the lowest points.

FIRE VENTILATION will provide a better condition for breathing and heat, but will not remove all the hazards or the dangerous gases.

ARTIFICIAL VENTILATION at both the floor and ceiling will provide greater safety, when removing extensive amounts of flammable vapors.

VENTILATE AS DIRECTLY OVER THE FIRE as possible is the best rule of thumb for selecting the point to open the roof.

Avoid cutting several small holes when VENTILATING.

Avoid cutting a hole larger than 8 FEET BY 12 FEET when VENTILATING through the roof.

For VENTILATING a roof, the firefighter should cut a rectangular shaped hole.

When making a VENTILATION hole through the floor of a structure, drill hole near a beam before sawing.

Reasons for VENTILATION in Firefighting:

1. To save lives by eliminating smoke and gases which endanger the occupants of the structure, that may be trapped or unconscious.

2. Controls the spread of fire.

3. To discover the exact location of the fire by allowing the smoke to rise.

The most significant reasons for VENTILATION during Firefighting operations where an inside attack is considered, is that it will decrease the risk from HEAT and CARBON MONOXIDE.

At most structure fires the principle reason for VENTILATION is to allow Firefighters to advance hoselines close enough to the seat of the fire so it can be extinguished with a minimum of damage.

CROSS VENTILATION depends greatly upon the presence of cold air currents.

For FORCED AIR VENTILATION the opening for the replacement air should be equal to or larger than the opening for the venting hole.

VENTILATION of a building starts at the top and progressively downward.

It takes at least TWO OPENINGS to do an effective task of HORIZONTAL or LATERAL VENTILATION, because there must be an air inlet and an air outlet.

WINDWARD: the side of the building where the wind is hitting.

LEEWARD: the side of the building that is opposite the side where the wind is hitting.

WIND SPEED = measured in miles per hours or measured in meters per second.

Good CROSS VENTILATION in a building is best established by opening the leeward windows from the top first and the windward windows from the bottom.

In CROSS VENTILATION, when possible open double-hung windows on LEEWARD side first from top to bottom.

During VENTILATION operations:
1. Interior doors should be opened to allow air circulation to help remove smoke.
2. All covered openings on a level with or above the fire should be opened.
3. Openings on the roof will aid ventilation, even though the fire is on lower floors.

FORCIBLE ENTRY:

FORCIBLE ENTRY is the admission into a secured structure with a minimal amount of delay, frequently with the utilization of special tools.

FORCIBLE ENTRY TOOL are the devices transported on fire apparatus for the specific use of achieving entrance into structures or obstructions so as to perform firefighting and/or rescue operations.

When FORCIBLE ENTRY is to be used, first check to see if doors or windows are locked.

The pick point of a PICK-HEAD AXE will easily brake the glass of a tempered glass door, to allow ENTRY into a building.

The DETROIT DOOR OPENER can effectively be used to gain ENTRY through doors that swing inward.

The DETROIT DOOR OPENER can effectively be used in making an emergency hose clamp.

TEMPERED GLASS is less hazardous than regular glass.

BEFORE attempting to force any door the firefighters should:

1. Check to see if door is locked.
2. See if hinge pins can be removed.
3. Have hose lines available.

BREACHING: the opening of masonry walls.

FORCIBLE ENTRY: entry into a secured building with a minimum of delay, often by the use of special tools. (forcible entry tools).

FORCIBLE ENTRY TOOLS: tools carried on fire apparatus used to gain entry into buildings and obstructions so that firefighting and rescue operations may be carried out.

To prevent glass from sliding down an axe handle while a firefighter is breaking a plate glass window, the firefighter should STRIKE THE UPPER PART OF THE WINDOW FIRST/STANDING TO ONE SIDE.

The opening of ceilings from below, use PIKE POLE or PLASTER HOOKS.

AIR CHISEL is a FORCIBLE ENTRY device that is operated by the use of compressed air.

BATTERING RAM is a FORCIBLE ENTRY device that consist of a large beam with a larger head, sometimes made of metal, weighing about 50 to 60 pounds and equipped with handhold's. (used to knock down doors, walls, etc.)

FLAT HEAD AXE (floor axe) can be used as a FORCIBLE ENTRY TOOL, it is an axe with a flat head that can be used as a sledge hammer or similar device.

KELLY TOOL is a FORCIBLE ENTRY TOOL that is like a claw tool except the tool has an axe head or blade at one end and a forked blade on the other end.

K-TOOL is a FORCIBLE ENTRY TOOL that is used for the removal of cylinder locks.

LOCK-BREAKER used to open or break locks.

OVERHAUL-SALVAGE:

OVERHAUL: the last stage of firefighting, during which all of the fire is found and extinguished.

The principal reason for OVERHAUL OPERATIONS is to complete fire extinguishment.

During OVERHAUL:
1. Do not remove debris that may indicate arson.
2. Complete thorough extinguishment of fire.
3. Prevent rekindling of fire.
4. Leave fire area in a secure condition.

The least important basis for OVERHAUL is the preparation for the return of building occupants.

Start at the TOP and work downward, when OVERHAULING a multiple story structure.

Manpower is the determining factor as to how much OVERHAUL takes place while fighting the fire.

After a fire is extinguished, OVERHAUL should be executed by all firefighting personnel.

Before OVERHAUL OPERATIONS begin, it is important to make sure that a building is structurally safe.

When OVERHAULING a room with large amounts of stock, it is usually best to move the articles at the walls and work towards the center of the room.

OVERHAUL usually takes up more time than any other fire scene operations.

While Firefighters are performing OVERHAUL OPERATIONS, they should aware of evidence of arson.

POOR OVERHAUL OPERATIONS are the primary reason for REKINDLE of a fire.

SALVAGE is the process devised to lessen fire, smoke, water, exposure, and any other destruction before, during, and after fires or other incidents.

SALVAGE COMPANY is the unit within the Fire Department that specializes in SALVAGE OPERATIONS.

As far as salvage operations go, the Fire Departments RESPONSIBILITY is LIMITED after the premises have been released back to the owner.

SALVAGE COVERS are water repellant tarpaulin of cotton, plastic, or other materials.

SALVAGE COVERS are utilized for the protection of property from water, smoke, and slight heat damage after or even during a fire.

The ACCURATE APPLICATION OF WATER is the most significant of salvage principles.

To protect contents of a building and reduce water damage during a fire, use SALVAGE COVERS over as much of the building and its contents as possible.

CARRYALL: salvage device that is 6 feet or eight feet square with rope handles at the edges. (used to carry or catch debris).

FLOOR RUNNER: usually 3 feet by 18 feet (can be up to 30 feet long). Made of canvas or plastic and used on floors to prevent damage to the floor or carpeting from mud, debris, or water.

SALVAGE OPERATIONS: technique used to reduce fire, smoke, water exposure, and other damage before, during, and after a fire.

SALVAGE basically is for the protection of building contents from water and to contain and discharge the water to the outside of the structure.

SALVAGE OPERATIONS do not have to wait for ventilation to be completed.

If manpower allows, SALVAGE OPERATIONS should be started as soon as the location of the fire has been discovered.

If manpower is low, SALVAGE OPERATIONS may have to be put off until hose lines are laid.

"COVER WORK" should start as soon as possible whether the fire is out or not, on the floor directly below the fire floor.

To reduce water damage, use SALVAGE COVERS over as much of the building and its contents as possible.

In a multistory structure fire, SALVAGE COVERS should be put in service over articles near the fire and on floors below the fire.

SALVAGE OPERATIONS should start on the fire floor and proceed downward.

The most sensible way to decrease water damage in a structure fire is to INCREASE THE VAPORIZATION of water used in the extinguishment.

EXPLOSIONS:

EXPLOSION: a immediate fierce liberation of energy from a substance or mixture as it decomposes, or undergoes swift chemical reaction, or changes from solid to a liquid.

EXPLOSIVE : any chemical compound mixture, or mechanism, with the principal or routine objective is to create an explosion.

BRISANCE is the effect of shattering, explosive (measurement).

CHEMICAL REACTIONS double their rate with each 18% RISE in temperature.

FLAMMABLE EXPLOSIVE LIMITS have upper and lower limits expressed in percentage of vapor within air.

Most intense explosion will occur with vapor at MID-RANGE of products explosive range.

All factors being equal, most violent explosions of vapor-air mixtures occur at the INTERMEDIATE LEVEL. (mid-range).

The factor that primarily determines the degree of destructiveness of a DEFLAGRATION is the rate of pressure rise.

BLEVE: Boiling Liquid Expanding Vapor Explosion.

BLEVE SIGNS:

1. Direct heat on vessel.

2. Rising sound of vessels relief valve.

BLEVE: failure of a major container into two or more pieces, at a moment when the contained liquid is at a temperature well above its boiling point at (normal atmospheric pressure).

The WIDER the flammable range of a gas then the more dangerous the gas.

DUST CLOUDS of most metals are explosive.

As dust particle size DECREASES the explosion potential increases.

Impurities REDUCE the potential of dust explosion.

Lowering the Oxygen content REDUCES the potential of dust explosion.

DUST EXPLOSIONS take place when dust of particular substances, correctly distributed, is ignited by a source of heat.

The FLAMMABLE or EXPLOSIVE LIMITS of a liquid are the minimum and maximum concentrations of vapors and are designated in terms of percentage by the volume of gas or vapor in oxygen.

EXPLOSIVE RANGE: the concentrations of upper and lower or flammable limits of a substance or mixture of substances, represented in expressions of percentage of vapor or gas in the air by volume.

WATER SUPPLY

The RATIO of WATER surface to GROUND surface on the earth is about 3 TO 1.

BASIC COMPONENTS OF A WATER SYSTEM:

1. Processing or treatment facilities.
2. Distribution system, including storage.
3. Source of supply.
4. Mechanical or other means of moving water.

THE THREE CLASSES OF WATER MAINS ARE REFERRED TO:

1. Primary feeders.
2. Secondary feeders.
3. Distributor mains.

The ISO recommends the SECONDARY FEEDERS be spaced not more than 3000 FEET apart.

PRIMARY FEEDERS: are large pipes used for moving water from the source, supply or storage area, to the secondary feeders. Primary feeders vary in size from 60 inches in diameter in large cities to 12 inches in diameter in small cities.

SECONDARY FEEDERS: are larger than distribution mains used in grid systems but are smaller than primary feeders. The function of the secondary feeders is to reinforce the distribution system. Secondary feeders tie the grid system to the primary feeders so as to aid in the concentration of the required fire flow at any point within the grid network.

SECONDARY FEEDERS should be LOOPED, which increases the capacity and the reliability of the entire system.

DISTRIBUTION MAINS supply the water directly to the fire hydrants and occupancies for domestic use.

DISTRIBUTION MAINS are cross connected so as to supply a grid throughout the built-up areas of the community. Water should be supplied to the hydrants from two directions (looping). Looping doubles the capacity and enhances the reliability.

DISTRIBUTION MAINS should be 12 inches in diameter on principle streets, with 8 inch diameter mains cross-connected every 600 feet in business districts and 6 inch mains in residential areas, cross-connected no more than 600 feet apart.

The FIRST consideration for water supply system is the DISTRIBUTION SYSTEM.

DISTRIBUTORS in a water distribution system: the section of a gridiron arrangement of small mains that facilitate individual hydrants and blocks of customers.

Mains in a GRIDIRON allowing water to flow from several directions to a point will cut down friction loss so as to insure a DEPENDABLE supply of water.

DIRECT PUMPING WATER SYSTEM: water is pumped directly into the distribution system from an elevated storage.

As far as providing water for municipal areas, the water supply is divided into ADEQUACY and RELIABILITY.

MOST RELIABLE type of water supply system is the GRAVITY, from impounding reservoirs.(combo gravity-pumping systems are the most common

The most reliable type of water system for fire protection of a City of a population of 500,000 or more is a supply system obtained from IMPOUNDING RESERVOIRS and brought to the City by GRAVITY.

CISTERN: water storage repository, normally below grade.(most dependable type of static water supplies = CISTERNS and GROUND TANKS)

WATER SUPPLY: the amount of water obtainable in an district for firefighting.

WATER SUPPLY POINT: a source of water in adequate supply for firefighting.

WATER TABLE: the upper confines of the part of the ground that is completely permeated with water.

To compensate for unreliable characteristics of a water system, store water in suitable amounts to dispense for fire protection and domestic needs in ELEVATED RESERVOIRS.

For fire or domestic use, for economical and normal practice that PUBLIC water systems be a SINGLE system to facilitate both needs.

WATER CONSUMPTION-REQUIREMENTS:

WATER CONSUMPTION: The quantity of water typically used by a district during a designated period of time, usually represented on a per capita basis or in millions of gallons per day.

WATER CONSUMPTION FOR FIRE USE: the entire amount of water utilized to extinguish a fire, determined by multiplying the rate and time of the fire flow.

TYPES OF WATER CONSUMPTION:

1. Average daily consumption.
2. Maximum daily consumption.
3. Peak hourly consumption.

AVERAGE DAILY CONSUMPTION = total amount of water used in a year divided by the number of days in the year.

MAXIMUM DAILY CONSUMPTION = the maximum amount of water used in a city during any 24 hour period in a 3 year period.

PEAK HOURLY CONSUMPTION = the maximum amount of water that can be expected to be used in any given hour of a day; usually = 2 to 4 times the normal hourly consumption.

WATER SYSTEM is considered ADEQUATE by ISO grading schedule when it can deliver the required fire flow for the required duration of hours, while domestic consumption is at its maximum daily rate.

WATER SUPPLY POINT: any source of water in suitable amounts for firefighting.

The value of a city's water supply system, for the fire service, mainly is dependent upon the ARRANGEMENT AND SIZE of the mains.

ORDINARY WATER SYSTEMS are designed for maximum working pressures of 150 PSI. Pressures of over 150 PSI are considered excessive.

FIRE FLOW: the amount of water accessible for fire fighting in a designated area.

FIRE FLOW is determined in addition to the normal water consumption in an area.

REQUIRED FIRE FLOW: is the amount of water needed for firefighting purposes in order to confine a major fire to the buildings within a block or other group complex.

MINIMUM FIRE FLOW is 500 GPM for un-congested areas of small dwellings.

MAXIMUM REQUIRED FIRE FLOW is 12,000 GPM for large industrial areas and for downtown areas of large cities.

ADDITIONAL FIRE FLOWS of from 2,000 GPM to 8,000 GPM are required for simultaneous fires.

TOTAL MAXIMUM REQUIRED FIRE FLOW for large areas, industrial and downtown areas of large cities is 20,000 GPM.

WATER SYSTEM is considered RELIABLE when it can supply the required fire flow for the number of required duration hours, with the domestic daily rate under certain emergency or unusual conditions at the maximum daily rate.

The CAPACITY of a PUBLIC WATER SYSTEM is determined by the total amount of water it must furnish divided by the water required for domestic and industrial use plus the water required for the fire service.

FIRE FLOW:

FIRE FLOW is normally thought of as the quantity of water available for firefighting in a given area.

FIRE FLOW is calculated in addition to the normal water consumption in the area considered.

The term FIRE FLOW is often used to describe the rate of application of an agent that a given fire requires.

FIRE FLOW may also be used to describe the rate of application of dry chemicals, powders, foam, and gases.

FIRE FLOW IN A CITY IS DEPENDENT UPON:

1. The size of the most congested area.
2. The hazards of the most congested area.
3. The structural conditions of the most congested area.

Fire Departments base their FIRE FLOW requirements on their need to furnish hose streams between 250-300 GPM EACH.

HYDRANTS:

HYDRANT: a casting of metal connected to the water system and is equipped with one or more valved outlets for hose or pump suction to be connected.

HYDRANT DESIGNATIONS:

1. Class A = green color rated @ 1000 GPM+.
2. Class B = orange color rated @ 500-1000 GPM
3. Class C = red color rated less than 500 GPM

HYDRANT CAPACITIES are rated by flow measurements and test of individual hydrants at a period of ordinary demand. The rating is based on 20 PSI residual pressure, when initial pressures are over 40 PSI (static). When initial pressures (static) are less than 40 PSI, residual pressures shall be at least 1/2 the initial rate.

NORMAL hydrant flow pressures are 65 PSI to 75 PSI.

MINIMUM FIRE FLOW PRESSURES from a hydrant (desirable) is 20 PSI, for large mains or well distributed hydrants it is 10 PSI.

Hydrants should be **INSPECTED SEMIANNUALLY** and **AFTER USE.** (preferably in the spring and fall)

LAMINER refers to very little turbulence.

RESIDUAL PRESSURES from hydrant to pumper:

1. Minimum of 5 PSI to 10 PSI, if hose between pump and hydrant is adequate to provide the desired flow at these residual pressures.
2. Desirable minimum pressure is 20 PSI for proper operation of pumpers.

RESIDUAL PRESSURE: the pressure remaining on the inlet side of a pumper, water main, or any water system, while water is flowing.

When **FIRE HYDRANTS** are spaced too far apart the effective capacity of a pumper will be reduced, because of the higher pressure that it must deliver.

SUPPLEMENTARY WATER SUPPLIES FOR A PUMPER:

1. Relay from other pumpers, on mains that are not affected by the fire flow.
2. Suction from static sources.
3. Large tank trucks.

STANDARDS for hydrants are prepared by "AMERICAN WATER WORKS ASSOCIATION".

HIGH pressure hydrant is 150 TO 300 PSI.

A rough rule to follow when HYDRANTS are to be positioned, is to place one hydrant near each intersection and an additional one between, if the distance is over 350 FEET TO 400 FEET MAXIMUM.

HYDRANT VALVES open against the pressure and close with the pressure.

Most fire hydrant valves turn COUNTERCLOCKWISE to open and CLOCKWISE to close.

DRY HYDRANTS do not have any positive pressure water source, they are like permanently installed hard suction to other type water source.

INSPECT hydrants twice a year.

NON-INDICATING VALVE = valve buried and which operates with a special wrench.

INDICATING VALVE:

1. OS&Y valve.
2. PIV.

WATER PIPES-MAINS:

MINIMUM DIAMETER of water main in residential area is 6 inches.

An 8 inch main has TWICE the capacity of a 6 inch main.

TUBERCULATION: deposits that accumulate in water mains.

TURBULENCE has greatest affect on loss of pressure on flowing water in pipe. (caused by tuberulations, roughness in the pipe wall).

CARRYING CAPACITY of WATER MAINS are relative to the size of the water main:

1. DOUBLING the size of the pipe, you increase the capacity by 6 TIMES.
2. TRIPLING the size of the pipe, you increase the capacity by 18 TIMES.
3. QUADRUPLE size of pipe, you increase the capacity by 38 TIMES.
4. QUINTUPLE size of pipe, you increase the capacity by 69 TIMES.

THERMOSYPHON = flow characteristics.

WATERWAY = the internal passage for the flow of water in a hose, pump, pipe-way, or other types of equipment.

WATER:

WATER (H_2O): a colorless, odorless, tasteless liquid that is widespread in its use for the extinguishment of fire.

WATERS existing forms are:

1. Liquid.
2. Solid = ice.(allotropic)
3. Vapor = steam.(allotropic)

WATERS freezing point is at 32 degrees F and the boiling point is at 212 degrees F.

WATER is useful as a cooling instrument because it has the greatest latent heat of vaporization of all the common materials = 970 Btu per pound.

WATER as a COOLING AGENT for flammable liquids:

1. Cuts off release of vapor from the surface of a high flash point oil, thus extinguishes the fire.

2. Protects firefighters from flame and radiant heat.

3. Protects flame exposed surfaces.

WATER as a MECHANICAL TOOL for flammable liquids:

1. Controls leaks.

2. Directs the flow of the product to prevent it's ignition, or to move the fire to an area where it will do less damage.

WATER as a DISPATCHING MEDIUM for flammable liquids:

1. Will float oil above a leak in a tank or during a fire.

2. Will cut off the fuels escape route by pumping it into a leaking pipe before the leak.

WATER may be used to smother flammable or combustible liquid fires when the liquid has:

1. Flash point above 100 degrees F.

2. Specific gravity of 1.1 or heavier.

3. When liquid is not water soluble.

WATER is the most important extinguishing agent because of its physical characteristics, its universal availability, and because of it's low cost.

WATER'S HIGH SPECIFIC HEAT is one reason that it has the ability to extinguish fires by cooling them below their ignition temperature.

WATER will absorb heat to a much greater extent than any other material that is easily available.

WATER has a lack of opacity, thus it has little ability to prevent the passage of radiant heat.

When WATER changes from LIQUID TO VAPOR it absorbs 10 TIMES the heat as any other extinguishing agent.

NATURAL LAWS THAT AFFECT THE USE OF WATER AS AN EXTINGUISHING AGENT:

1. The Law of Heat Flow.

2. The Law of Specific Heat.

3. The Law of Latent Heat of Vaporization.

Compared with spray streams, solid streams generally are MORE EFFECTIVE in extinguishing sub-surface fires, such as deeply charred wood.

THEORETICALLY, 1 gallon of water at 90 degrees F. cooling capacity will cool 100 cubic feet of fire within 30 seconds.

WATER FOG is not capable of extinguishing flammable liquid fires with a flash point below 100 degrees F

SPRAY FIRE STREAMS IN CONTRAST TO SOLID STREAMS:
ADVANTAGES:

1. Absorbs more heat, more rapidly.

2. Covers a greater area with water.

3. Uses less water.

DISADVANTAGES:

1. Requires higher discharge pressure.

2. Has a shorter reach.

3. Has less penetration.

4. Has less cooling effect in subsurface areas such as charred wood, etc.

A gallon of water produces a maximum of 200 CUBIC FEET OF STEAM.

The formation of steam by water that has been applied to the burning source will temporarily cause an INERT gaseous water zone (steam) in and around the burning zone.

EXTINGUISHING SYSTEMS

FIRE EXTINGUISHER:

FIRE EXTINGUISHER: a portable hand-held device or device on wheels that is designed and used to extinguish small fires.

FIRE EXTINGUISHER RATINGS AND COLOR CODES:

Class A fires = Ordinary combustibles:
   Green triangle.

Class B fires = Flammable liquids:
   Red square.

Class C fires = Electrical:
   Blue circle.

Class D fires = Metal:
   Yellow star.

Portable fire extinguisher are classified according to their INTENDED USE.

2A WATER fire extinguisher has a 100 FOOT COVER area.

The stream range of a 2A WATER 2 1/2 GALLON loaded stream extinguisher will discharge an effective stream for about ONE MINUTE.

REGULAR DRY CHEMICAL EXTINGUISHER are Bicarbonate base powder for Class B and Class C fires. (Sodium Bicarbonate and Potassium Bicarbonate)

MULTIPURPOSE DRY CHEMICAL EXTINGUISHER are Ammonium Phosphate base powders for Class A, B, and C fires. (Monoammonium Phosphate, Barium Sulfate)

DRY CHEMICAL EXTINGUISHER are a mixture of Sodium Bicarbonate, Potassium Bicarbonate, or Ammonium Phosphate.

MULTIPURPOSE DRY CHEMICAL EXTINGUISHER, extinguish by the Ammonium Phosphate decomposing and leaving a sticky residue on the burning material, this seals the material off from Oxygen.

DRY CHEMICAL EXTINGUISHER EXTINGUISH BY:
1. Smothering.
2. Cooling.
3. Radiation shielding.
4. Chain breaking; a reaction in the flame may be the principle cause of extinguishment.

DRY CHEMICAL EXTINGUISHER should not be used where relays or other delicate electrical contacts are located such as telephone exchanges.

DRY CHEMICAL EXTINGUISHER will not extinguish fires in materials that supply their own Oxygen.

15 LBS of DRY CHEMICAL in an extinguisher has a maximum effective range of about 12 FEET.

CARBON DIOXIDE FIRE EXTINGUISHER extinguish by cooling with the rapid expansion of liquid to gas producing a refrigerating effect.

CARBON DIOXIDE EXTINGUISHER with CO_2 under pressure, the purpose of the horn dispenser is to avoid entraining air as the contents pass through the small orifice at high velocity.

CARBON DIOXIDE EXTINGUISHER suppress chain reaction of combustion.

Firemen should exercise care when using CARBON DIOXIDE EXTINGUISHER, mainly because of the possibility of RE-FLASH.

15 LBS CARBON DIOXIDE EXTINGUISHER range = 8'-10'.

CARBON TETRACHLORIDE vaporizes to form a heavy non-flammable gas.

PORTABLE FIRE EXTINGUISHER: an easily moved tool which holds chemicals, fluids, or gases that will extinguish fires.

EXTINGUISHMENT BY SEPARATION cannot be accomplished with burning materials that contain their own oxygen supply, such as Cellose Nitrate.

PORTABLE FIRE EXTINGUISHER WHICH ARE CONSIDERED OBSOLETE:

1. Soda Acid.
2. Carbon Tetrachloride.
3. Loaded Stream.
4. Cartridge-operated Water.
5. Inverting Foam.

EXTINGUISHMENT BY SEPARATING the oxidizing agent from the fuel is accomplished by blanketing or smothering a fire.

CHEMICAL COMPOSITION of most ordinary combustible solids consist of primarily: Carbon, Hydrogen, and Oxygen.

COMBUSTIBLE METAL FIRES (Class D) are extinguished by a number of approved extinguishing agents; Powders, and Dry Powders. Each for specific metals.

AFFF: AQUEOUS FILM FORMING FOAM, is a fluorinated foaming agent that creates a film which floats on flammable liquids or water to exclude Oxygen and thus smother the flames of the fire.

The notion of the FOAM EXTINGUISHER is that the Oxygen is excluded by the bubbles, and also the surface is cooled.

EXTINGUISHING AGENT: any material which will extinguish a fire by cooling the ignited substance or by obstructing the supply of oxygen, or by chemically suppressing combustion.

HALOGEN: any one of the five elements which manufacture salt when blended with metals, which are fluorine, chlorine, bromine, iodine, and astatine.

HALOGENATED AGENT: is an extinguishing agent made up of hydrocarbons that have one or more hydrogen atom that has been replaced by HALOGEN atoms; the standard

HALON EXTINGUISHER: is charged with a halogenated agent and an expellant. Used on flammable liquid fires and charged electrical equipment fires.

HALOGEN elements used are fluorine, chlorine, bromine, and iodine.

HALON: is one of many HALOGENATED hydrocarbon mixtures, two of them that are a standard as extinguishing agents (bromotrifluormethane and bromochlorodifluoromethane); these two are inert to virtually all chemicals, and resistive to high and low temperatures.

HALOGENATE EXTINGUISHING agents work by:

1. Vaporizing liquid fire extinguishing agents.
2. Chain Breaking.
3. By providing non-flammability and extinguishing characteristics.

HALON EXTINGUISHING AGENTS:

1. Fluorine.
2. Chlorine.
3. Bromine.
4. Iodine.

WETTING AGENTS-FOAM:

Adding WETTING AGENTS to water increases the waters heat absorption ability.

WETTING AGENTS increase the rate of heat absorption.

WET WATER creates a problem, because the effectiveness of salvage covers is reduced.

The main effect of adding a WETTING AGENT to water, normally is that it increases the rate of heat absorption.

WETTING AGENT: is a chemical additive that is used to change water and assist for greater penetration by reducing the surface tension of the water.

WETTING AGENTS increase the waters penetration ability.

WETTING AGENTS reduce the surface tension of water.

WET-WATER solutions will foam easily, and the temporary foam will control and extinguish class B fires better than ordinary water.

Water which has had various types of CHEMICAL AGENTS (wetting agents) added WILL RETAIN all its basic qualities with the addition of the capabilities of wetting, spreading, and penetration much more quickly into materials.

SURFACE TENSION is the resistance to penetration possessed by the surface of a liquid.

WETTING AGENT foams breakdown to original liquid form at temperatures below boiling point of water, and retains its wetting and penetrating properties which aid in cooling and extinguishing the fire.

WETTING AGENT in soda acid extinguisher will decrease the range and cause a spraying stream.

VISCOUS WATER: water that includes a thickening agent to diminish surface runoff.

VISCOUS WATER (thickened water):
Advantages:

1. Sticks to burning fuel.

2. Spreads itself in continuous coating.

3. Thicker than plain water.

4. Absorbs more heat.

5. Projects longer and higher straight streams

6. Seals fuel from oxygen after drying.

7. Resist wind drift. (as from aircraft in forest fires).

LIGHT WATER (fluorinated surfactant), is useful in obtaining quick knockdown of flammable liquid fires, and in providing a vapor sealing effect for reducing subsequent flashover of fuel vapors exposed to lingering.

LIGHT WATER = AFFF (Aqueous Film Forming Foam); may be used concurrently with dry chemicals to extinguish flammable liquid fires by smothering them.

BASIC TYPES OF FOAM FOR FIREFIGHTING:

1. Chemical foam.
2. Mechanical foam.

SUBSURFACE FOAMS injected or intermixed with the liquids that are involved are called flouro-protein.

AFFF FOAM RATIO:

1. Hydrocarbons (petroleum products) = 3% foam to 97% water.
2. Polarsolvents (water soluble) = 6% foam to 94% water.

FOAM PROPORTIONER: a device used to blend a foam concentrate with water while forming air foam.

FOAM should always be applied to oil fires gently.

FOAM is the most effective extinguishing agent for oil fires, because it is lighter than oil and remains on top.

CONVENTIONAL FOAM is formed by the reaction of alkaline salt solution in acid salt solution in the presence of a foam stabilizing agent, and mechanical or air foam formed by turbulent mixing of air with water containing foam forming agents.

ORDINARY FOAM IS BROKEN DOWN BY:

1. Common alcohols.
2. Aldehydes.
3. Ethers.

ALCOHOL FOAM is recommended for all water soluble flammable liquids except for those that are only very slightly soluble.

SPRINKLER SYSTEMS:

SPRINKLER SYSTEM: the system of water pipes that are spread with sprinkler heads and located in/on a structure so as to contain and extinguish fires.

SPRINKLER SYSTEM: use appropriate water supplies, such as gravity tanks, fire pumps, reservoirs, pressure tanks, or by connecting to the city water mains.

SPRINKLER SYSTEMS: normally have controlling valves along with an alarm that will signal when the system is activated.

AUTOMATIC SPRINKLERS should be installed in all buildings that, by reason of their size, construction, or occupancy, constitute a serious life hazard or that may be a conflagration breeder.

The main principle in the control of fire using AUTOMATIC SPRINKLERS, especially where ordinary materials are involved, is to detect the fire in its incipient stage and to drench the immediate fire area so as to limit fire spread and damage to property.

AUTOMATIC SPRINKLERS, regardless of type, are designed to distribute water upon a fire automatically and in sufficient quantities to extinguish or control the fire until firefighting units arrive.

AUTOMATIC SPRINKLERS can provide a cooling effect that is greater than the heat generated by the fire, thus the sprinklers can gain control.

TYPES OF AUTOMATIC SPRINKLER SYSTEMS:

1. Wet pipe system.
2. Dry pipe system.
3. Pre action system.
4. Combined system.
5. Deluge system.
6. Limited water supply system.
7. Partial system.

The success of AUTOMATIC SPRINKLER application is dependent upon an adequate water supply.

SPRINKLER SUPERVISORY SYSTEM = monitored system.

When fire officers discuss the first line of defense against the spread of fire, they will be discussing AUTOMATIC SPRINKLER SYSTEMS.

SPRINKLER SYSTEM: usually will have a controlling valve along with an alarm that will signal when the system is activated.

SPRINKLER CONNECTION: a connection that hose lines from a fire pumper are used to increase the pressure within the system. (usually a sprinkler siamese with two inlet connections).

SPRINKLER RATINGS:

Ordinary	- No color	= 135 - 170 degrees F.
Intermediate	- White	= 175 - 225 degrees F.
High	- Blue	= 250 - 300 degrees F.
Extra high	- Red	= 325 - 375 degrees F.
Very extra high	- Green	= 400 - 475 degrees F.
Ultra high	- Orange	= 500 - 575 degrees F.

SPRINKLER COVERAGE according to occupancy hazard, with 1/2 inch orifice:

Light	= 130 square feet to 168 square feet.
Ordinary	= 130 square feet.
Extra high	= 90 square feet.

AUTOMATIC SPRINKLERS should be installed in warehouses of Type I or II, but not of Type III.

AUTOMATIC SPRINKLERS are the most effective safeguard against loss of life by fire. Psychological as well as physical, by minimizing panic.

PRE-ACTION SPRINKLER SYSTEMS are activated at the water supply valve, not at the sprinkler head, like in the standard type dry sprinkler system.

WET PIPE SPRINKLER SYSTEM fully charged with water.

DRY PIPE SPRINKLER SYSTEMS, the piping contains air under pressure, 15 PSI to 20 PSI and no water in the piping.

In a DRY PIPE SYSTEM no water passes the valve in the riser, air is pressurized in the system.

Advantages of a PRE-ACTION SYSTEM:

1. Protects property from damage that would otherwise be caused by a broken pipe or sprinkler head.

2. Allows water to flow into the piping system before the heads open, thereby reducing the "delay time" normally associated with the Dry pipe systems.

3. May be supervised by an appropriate alarm system.

PRE-ACTION DELUGE: dry system with open deluge heads.

PRE-ACTION SPRINKLER SYSTEMS ARE ACTIVATED BY:

1. Smoke detectors.

2. Heat sensors.

The typical PRE ACTION SYSTEM is basically a dry pipe system and uses heat detecting devices that open the water supply valve, sending water into the system and to the sprinkler heads whether they are open or closed.

COMBINED SYSTEMS are pre action and dry pipe systems in which a heat responsive device operates a dry pipe valve before any sprinklers are opened.

DELUGE SPRINKLER SYSTEMS wet down the entire area by admitting water to sprinklers that are open at all times.

DELUGE SPRINKLER SYSTEM should be used in places where ceilings are very high and the drafts would prevent an ordinary system from operating properly.

DELUGE SYSTEMS use only sprinkler heads of the open orifice type.

DELUGE SYSTEMS should be used only where large amounts of water are required immediately, such as in areas where flammable's are stored.

LIMITED WATER SUPPLY SYSTEMS are used where the water supply is limited, these special systems are similar to a normal wet pipe system, except that the amount of water available is minimal.

PARTIAL SYSTEM are used when only a certain area in a building requires protection, such as stairways and corridors.

OUTSIDE SPRINKLER SYSTEMS, the water curtain is for exposed protection.

SPRINKLER SYSTEMS SHOULD BE PUMPED AT 150 PSI WHEN:

1. Fire and or smoke showing.
2. Water in system is flowing.

SPRINKLER SYSTEMS should be pumped at a MINIMUM OF 100 PSI.

WATER SUPPLIES FOR SPRINKLER SYSTEMS:

1. Public water works systems.
2. Public and private supplies.
3. Gravity tanks. (minimum of 5000 gallons)
4. Pressure tanks. (minimum of 4500 gallons)
5. Fire pumpers.
6. Fire department connections.

OS&Y VALVE:

1. Outside Screw and Yoke Valve.
2. Outside Stem and Yoke Valve.

PIV = POST INDICATOR VALVE.

STANDPIPE SYSTEMS:

Class I, Class II and Class III STANDPIPES: SHALL all maintain a residual pressure of 65 PSI at the top most outlet.

CLASS I STANDPIPE SYSTEM:
For use by Fire Departments and those trained in handling heavy fire streams. (2 1/2 inch hose inside building or for exposure.)

CLASS II STANDPIPE SYSTEMS:
For use primarily by the building occupants until the Fire Department arrives. (1 1/2 inch hose for incipient fires.)

CLASS III STANDPIPE SYSTEMS:
For use by either Fire departments and those trained in handling heavy hose streams or by the building occupants on small hose streams. (for large or small fires.)

TYPES OF STANDPIPE SYSTEMS:

1. Wet, supply valve open with water pressure at all times.
2. Dry, no permanent water supply in sprinklers.
3. Automatic supply system, opening hose valve.
4. Manual to remote system, is at hose station.

When fighting a fire in a structure that has a STANDPIPE SYSTEM, the system should be used instead of advancing hose on fires above the 3RD FLOOR.

Use the STANDPIPE SYSTEM instead of hose lines if the fire cannot be reached with pre-connected lines from the pumper.

A fire pumper cannot pressurize a STANDPIPE SYSTEM through the siamese connection until the engine pressure is greater than the pressure at the check valve of the system.

If the STANDPIPE SIAMESE connection (inlet) at ground level is damaged, charge the system with hose lines on the 1ST FLOOR outlets.

When a structure is protected by both STANDPIPE and SPRINKLER SYSTEMS, the Fire Department should charge the STANDPIPE SYSTEM FIRST.

IN HIGH RISE STANDPIPE SYSTEMS, there should be tanks or pumps every 20 stories.

According to the ISO, GRAVITY water systems are ALWAYS more acceptable than pump systems.

FIRE HOSE AND STREAMS:

N.F.P.A. #196 is the standard for fire hose.

N.F.P.A. #198 is the standard for the CARE of fire hose.

Each ENGINE COMPANY should have at least 2400 FEET of 2 1/2" hose or larger.(could be 2 1/2" and 4" hose). 2400 feet = 1200 on the apparatus and 1200 feet in storage.

Specifications for fire apparatus HOSE COMPARTMENTS are found in NFPA 1901.

Hose compartments on fire apparatus are usually called HOSE BEDS.

The divider or separator in a fire hose compartment is called BAFFLE-BOARD.

THREE BASIC CONSTRUCTION METHODS OF FIRE HOSE:

1. Braided   2. Wrapped   3. Woven

Jackets of multi-jacket hose may be SEPARATE and INTERWOVEN.

UNLINED FIRE HOSE is usually used inside buildings on standpipe risers.

The FEMALE COUPLING should be connected to the pump discharge for pre-connected hoselines.

For hose lays from the water source to the fire should have the hose loaded so that the FEMALE COUPLING will come off first.

HOSE LOAD = total hose carried on apparatus.

ADVANTAGES OF THE DONUT-ROLL:

1. Both ends are available on the outside of the roll.

2. The hose is less likely to spiral or kink when unrolled.

When loading hose and a coupling is in the wrong place, the fold to use is called a DUTCHMAN.

The most important factor concerning the life of fire hose is the care that the fire hose receives AFTER FIRES.

RECOMMENDED PRACTICES FOR MINIMIZING FIRE HOSE INJURY DUE TO HEAT:

1. Protect hose from excessive heat or fire whenever possible.

2. Use a current of warm air for drying hose.

3. Do not dry fire hose on hot pavement.

DEFECTIVE hose stream caused by:

1. Too much pressure.

2. Too little pressure.

3. Air in line.

4. Kinks in hose.

5. Hose twisted near nozzle.

6. Defective nozzle.

Hose should cross streets OPPOSITE the fire building.

With constant engine pressure on a hose lay of moderate length, if you use a SMALLER nozzle you will increase the nozzle pressure and increase the range or the reach of the stream.

Fire hose should have an expected life of about TEN YEARS.

When hose bursts under pressure it can start to whip around; the best place for a firefighter to be is INSIDE the bend.

FIRE HOSE MUST:

1. Withstand high pressures.
2. Transport water with a minimum loss in the working pressure.
3. Be flexible enough so that it can be handled and used under extreme fire conditions without requiring a large number of firefighters.

The BEST time to use a booster line is for quick attacks on a fire in its incipient stage.

HEAT is the cause for rubber lined hose to be damaged most rapidly.

RUN WATER through rubber lined hose at least every 6 months.

Store hose loosely rolled in a cool dry place.

If you must drive over fire hose, it is best to use a HOSE BRIDGE.

If necessary to drive apparatus over hose lines during the firefighting operations, one rule which applies under all conditions is that: apparatus drive wheels should COAST SLOWLY over hose lines.

HOSE BRIDGE: a wedged shaped ramp used to allow vehicles to pass over fire hose.

If you must drive over fire hose WITHOUT the use of a hose bridge, the hose should be charged with water under pressure.

As a general rule, the diameter of a plain nozzle to be used on a line should NOT EXCEED 1/2 of the diameter of the hose.

DOUGHNUT ROLL: 50' length of hose doubled and rolled toward coupling.

MINIMUM amount of reserve hose = one full load.

Amount of hose needed for FIRE BUILDING = 1 length per story plus one length.

2 1/2" FIRE HOSE is CONSIDERED:

1. SHORT LAY if = 0' to 300' long.
   (Maximum tip = 1 1/4")

2. MEDIUM LAY if = 300' to 600' long.
   (Maximum tip = 1 1/8")

3. LONG STRETCH if = 600' to 900' long.
   (Maximum tip = 1")

Additional lines should not be connected to the pumper if the intake pressure is below 5 to 10 PSI unless the NOZZLE TIPS already in use are REDUCED.

MAXIMUM DIAMETER NOZZLE TIP SIZE FOR:

1. SHORT LAY=1/2 the hose diameter.

2. MEDIUM LAY=1st size smaller than 1/2 hose diameter.

3. LONG STRETCH=2nd size smaller than 1/2 hose diameter.

Each ENGINE COMPANY should carry:

1. 1200 feet of 2 1/2 inch hose or longer.

2. 400 feet of 1 1/2 inch hose.

3. 200 feet of 1 inch hose.

DOUBLE JACKET hose will permit minimum twisting.

DOUBLE JACKET hose's greatest advantage over single jacket hose is durability.

Main reason for double jackets on fire hose is to allow HIGHER working pressures.

COTTON JACKET hose is the most susceptible hose to mildew.

The ability of fire hose to withstand the high pressures they undergo, without bursting, is because of the COTTON JACKET.

Advantage of MULTIPLE JACKET HOSE over single jacket hose is that it has the ability to withstand a greater amount of chafing.

DACRON FILLER HOSE has greater friction loss than all cotton jacket fire hose.

The main advantage that DACRON fire hose has over cotton jacket hose is that it is LIGHTER in weight.

The best position for the nozzle when advancing dry hose is OVER THE LEFT SHOULDER WITH THE NOZZLE IN THE BACK OF THE BODY.

A line of fire hose from which water is flowing through a nozzle, an open butt, or a broken line, and which is not under control by the firefighters is called A WILD LINE.

The safest way to control a WILD LINE is by CLOSING A VALVE to shut off the flow of water.

One man can control 350 GPM TO 400 GPM flow with a good stream, pressure, and reach.

When advancing hose lines up a ladder, the firefighters should have about 10 FEET between them.

Where fire hose is being advanced up a ladder there should be 20 - 25 FEET of hose between each firefighter.

Devices that are used with fire hose but water does not pass through are called HOSE TOOLS.

HIGBEE CUT is a tapering of a coupling thread to avert the cross threading of threads while making hose connections.

When a fire hose has been frozen into a thick sheet of ice, spreading SALT over the area can be effective in thawing the mass.

The most important aspect of picking up frozen hose after a fire is out = SPEED.

FIRE STREAM - EFFECTIVENESS - CAPACITIES:

On most PERIPHERY-DEFLECTED stream nozzles, the patterns are adjustable.

EXCESSIVE PRESSURE will break-up a small stream faster than a large stream.

Good fire stream for BOOSTER PUMP = 90 PSI to 100 PSI.

NON-CONSTANT GALLONAGE nozzles are adjustable and a change in the pattern will cause a change in the gallonage.

CONSTANT-GALLONAGE or CONSTANT-FLOW nozzles will discharge the same GPM regardless of the pattern.

ADJUSTABLE-GALLONAGE nozzle is a constant-flow nozzle in which the firefighters can choose the GPM.

FLOW PRESSURE: is the rate of flow of the water from a discharge opening.

If the nozzleman snaps the shut-off closed on a tip too rapidly, a high pressure surge (WATER HAMMER) could travel through the hose line back as far as the WATER MAIN.

CURVATURE OF A FIRE STREAM will offer the advantage of greater penetration into a structure.

Maximum stream PENETRATION ANGLE is at 45 DEGREES. If the angle increases the penetration decreases.

Greatest HORIZONTAL REACH occurs at 30 - 34 DEGREES angle.

Maximum effective VERTICAL REACH occurs at 60 - 75 DEGREES angle.

The efficiency of a FIRE STREAM depends upon the distribution of the water from the stream.

Water will vaporize into steam and be ineffectual if the FIRE STREAM is too small for the volume of heat generated by the fire.

A FIRE STREAM cannot be considered EFFECTIVE unless it is capable of carrying sufficient amounts of water so as to pass through the super heated gases and reach the burning substance.

Effective reach of a FIRE STREAM is effected by:

1. Nozzle size.

2. Nozzle pressure.

3. Air resistance.

The reduction of effectiveness of a large FIRE STREAM that is focused into higher windows of a high rise structure is primarily associated to the reduced building penetration because of the increased stream angle.

Hose that has been laid out in a straight line is charged with water the hose will SNAKE, this is due to the hose increasing in its length.

Too great of pressure on a HOSE STREAM with a small tip will cause the stream to break into a spray.

During fire situations hose lines should be laid in straight lines to avoid INTERMINGLING of hose lines from other fire units.

If two or more hose streams are advanced during a concentrated fire attack, the fog patterns should be INTERLOCKED to create a solid mass of water.

Fire department SOP should include that one or more protection lines should follow behind or beside advancing fire streams.

An EFFECTIVE FIRE STREAM discharges 90% of its volume inside a circle 15" diameter, and 75% of its volume inside a circle 10" in diameter.

If the rate of flow is adequate but the fire is not knocked down within 1 MINUTE, the seat of the fire is not being reached by the fire stream.

COEFFICIENT OF DISCHARGE = actual discharge from opening is a certain percent of the theoretical amount. Example : 90% = .9 as the coefficient.

Hand line MAXIMUM penetration distance = 50 feet into structure.

AIR IN HOSE LINES usually caused by operating pumps under negative pressure; this is objectionable because the air imprisoned in the hose stream tends to BREAK UP THE STREAM.

FORCE PUMPING: adding pressure sufficient to assure a stream velocity that will open the distance from the hose line to the fire.

COURSE of a fire stream is AFFECTED by:

1. Gravity.
2. Friction due to air resistance.
3. Wind velocity.
4. Obstacles.

Any WIND will hinder a fire stream.

Types of HOSE STREAMS: fog and straight; master streams and hand lines.

CHIEF PURPOSE of a nozzle on a fire stream is to give it shape and added velocity.

The PERFECT fire stream cannot be sharply defined.

Wind blowing directly into a fire stream will RAISE the VERTICAL REACH; SHORTEN the HORIZONTAL REACH.

BROKEN FIRE STREAM: stream of water that has been broken into coarsely divided drops.

MOBILITY of fire stream determines effectiveness.

3 1/2" hose carries more than TWICE the capacity of 2 1/2" hose.

A 2" NOZZLE TIP will provide a heavy stream of TWICE the capacity of an 1 1/2" NOZZLE TIP.

Hose of 1" I.D. will have 32 TIMES as much friction loss as hose of 2" I.D.

AS FAR AS FLOW, two 3" lines = three 2 1/2" lines:
1. 3" line   = 375 GPM;(2 X 375 = 750 GPM).
2. 2 1/2" line = 250 GPM;(3 X 250 = 750 GPM).

A single 2 1/2" hose line = 2/3 the FLOW of a single 3" hose line.

25,000 cubic feet is the maximum fire area for MANUALLY APPLIED HOSE STREAM.

MAXIMUM LENGTH for working hose lines= 400'to 500'.

MAXIMUM EFFICIENT water capacity of fire hose:
1. 1   " hose = 30 GPM.
2. 1 1/2" hose = 100 GPM.
3. 2 1/2" hose = 250 GPM.
4. 3   " hose = 500 GPM.
5. 3 1/2" hose = 750 GPM.

2 1/2" hose moves 1 TON of water per minute.

MAXIMUM VERTICAL REACH is attained when the nozzle is perpendicular to the ground, but is not used, maximum angle of common vertical stream is considered to be 60 - 70 DEGREE angle.

RECOMMENDED nozzle pressure for hand lines is 60 PSI to 70 PSI.

EFFECTIVE nozzle reach is at 60 PSI: The larger the tip the farther the horizontal reach. Not as much for vertical reach.

Heavy fire stream is one in EXCESS of 400 GPM.

FIRE STREAM: a stream of water from the time it leaves a nozzle until it reaches the point of intended use.

MASTER-SPRAY-FOG-STREAMS:

MASTER FIRE STREAM: any fire stream that is too large to be controlled without the use of mechanical help.

The terms "MASTER FIRE STREAMS" and "HEAVY FIRE STREAMS" are usually most closely related to the quantity of discharge.

MASTER STREAM DEVICES mounted on the deck of a fire apparatus and directly connected to the pump:

1. Turret pipe.

2. Deck gun or deck pipe.

Smallest tip for MASTER STREAM = 1 1/4".

MASTER FIRE STREAM: streams of water that are discharged from heavy stream appliances that use tips larger than 1 1/4". (in excess of 400 GPM).

MONITOR: a master stream device with the ability to change stream direction while water is flowing.

PENETRATION AND DEFLECTION: determine the effectiveness of master streams.

SPRAY for fire streams in CONTRAST to SOLID fire streams:

POSITIVE POINTS:

1. Absorbs more heat more rapidly.

2. Covers greater area with water.

3. Uses less water.

NEGATIVE POINTS:

1. Requires higher discharge pressure.

2. Has shorter reach.

3. Less penetration.

4. Less cooling effect in subsurface areas. (charred wood, etc.)

In general, the accepted effective FOG NOZZLE PRESSURE is at 100 PSI.

With adequate reach, the effectiveness of a fire stream, as far as fire extinguishing is concerned, increases the most as the stream takes on the characteristic of a FOG STREAM.

FOG NOZZLE for indirect firefighting, nozzle pressure = 100 PSI to 150 PSI.

For effective reach and penetration with the use of a FOG NOZZLE, operating pressures should = 50+ PSI.

Nozzle reaction on SPRAY NOZZLES is less than a straight stream because the smaller impinging streams will defuse the reaction forces.

FOG NOZZLES are preferred for fighting oil fires because the fog SMOTHERS the fire.

FOG STREAMS have a larger diameter than solid streams, therefore there is more area that has to deal with air friction loss, thus causing its forward velocity to be retarded more rapidly.

The SIZE of the fire area BEST determines the required reach of a fog stream.

COMMON DESIGNS OF NOZZLE CONTROL VALVES:
Rotary valves and Ball valves.

STREAM STRAIGHTENERS: are devices that prevent rotary motion and twisting of water currents at the nozzle so as to create a more perfect stream.

BRESNAN DISTRIBUTOR is designed to emit a heavy spray of water over an area of about 30 FEET.

DISTRIBUTORS THAT REVOLVE DURING OPERATION:
Bresnan and Gorter.

The pattern of a BRESNAN DISTRIBUTOR will have 75% of its volume directed DOWNWARD.

When a MYSTERY NOZZLE is operating at maximum capacity, the quantity of water is greater in the fog spray then in the solid stream position.

FOG STREAM: a billow of water mist discharged by a spray nozzle.

Methods used to deflect WATER JET so as to produce fog streams:

1. Periphery deflection.
2. Impinging deflection.
3. Whirling deflection.

SOLID STREAM: is a stream that is in a solid mass.

SPRAY STREAM: finely divided particles of water.

STRAIGHT STREAM: a solid stream of water used to gain maximum water force against an object or to maximize the water reach and penetration.

When a NOZZLE is CLOSED TOO QUICKLY:

1. Life and equipment are endangered.
2. Pressure becomes almost four times greater within hose lines and water mains.
3. Water mains may rupture.

NOZZLE PRESSURE: the velocity pressure in PSI at which water is discharged from a nozzle.

TESTING-FRICTION LOSS:

U.L. ACCEPTANCE TEST for DOUBLE JACKET FIRE HOSE:

1. Elongate 42 inches or less .
2. No more than 1 3/4 turns to the right. (tightens couplings)
3. Hydrostatic pressure of 400 PSI.

NEW DOUBLE JACKET hose tested at 400 PSI.

NEW SINGLE JACKET hose tested at 300 PSI.

U.L. STANDARD for minimum allowable BURSTING PRESSURE for double jacket hose is 600 PSI.

Hose line to be tested should not exceed 300 FEET.

COTTON DOUBLE JACKET HOSE should be TESTED ANNUALLY

1. At 250 PSI.
2. For 5 MINUTES.
3. In 300 FOOT lengths.

Upon receipt of NEW fire hose it should be tested at a pressure that complies with NFPA standard 196.

SERVICE testing of fire hose should comply with NFPA standard 198.

Fire hose SERVICE test is based on a minimum pressure of 250 PSI.

After proper pressure is reached during FIRE HOSE SERVICE TESTING, the pressure should be held for a minimum of FIVE MINUTES.

The maximum amount of time fire hose should go without a service test is ONE YEAR.

Fire hose should be capable of passing an annual service test at 250 PSI.

Fire hose should last a MINIMUM of 10 YEARS.

HARD SUCTION should withstand a working pressure of 200 PSI.

HARD SUCTION, when specified shall be re-enforced smooth bore rubber (designed for low friction loss) which will not collapse under vacuum of 23"HG and will withstand a pressure test of 200 PSI. Only expansion ring type couplings shall be used.

PRESSURE LOSS in unlined hose is about 2 TIMES that of lined hose.

When exceeding 250 PSI BOOSTER HOSE shall be rated at not less than 800 PSI working pressure. Booster line must be 200 feet long at 1 inch diameter).

HYDROKINETIC: water in motion; such as water flowing through a hose line.

Friction loss is governed by the QUANTITY (GPM) of water flowing.

Small suction hose RESTRICTS capacity of pumpers.

FRICTION LOSS VARIES:

1. Directly with the length.
2. Directly with the square of the velocity. (flow)
3. Inversely with the 5th power of the diameter.
4. Independently of the pressure.

Hose in zig-zag pattern will INCREASE friction loss by about 5% to 6%.

1 GPM is = 3.785 LITERS PER MINUTE.

FRICTION LOSS: when a dissipation of energy occurs during the flow of water in a fire hose.

FRICTION LOSS is caused by the turbulence of moving water against the walls of fire hose or water pipes.

FRICTION LOSS is expressed is PSI.

The FASTER that water is traveling forward, the FARTHER it will reach before being pulled to the ground by gravity.

To double flow at constant friction loss, pressure must be QUADRUPLED.

50 FOOT section of 2 1/2" cotton rubber lined hose, filled with water weighs 106 LBS.
(12.75 gallons X 8.35 LBS = 106 LBS)

1 FOOT section of 2 1/2" cotton rubber lined hose filled with water weighs 3 LBS to 4 LBS, counting the water and hose weight.

50 FOOT section of cotton rubber lined hose without water and not counting couplings weighs:

1. Single jacket = 35 LBS.
2. Double jacket = 50 LBS.

XXX

Editors note: this text includes excerpts from the following Books written by the author and available from Information Guides Dept. "B", P.O. Box 531, Hermosa Beach, CA 90254:

THE COMPLETE FIREFIGHTER CANDIDATE

FIREFIGHTER WRITTEN EXAM STUDY GUIDE

FIREFIGHTER ORAL EXAM STUDY GUIDE

FIRE ENGINEER WRITTEN EXAM STUDY GUIDE

FIRE ENGINEER ORAL EXAM STUDY GUIDE

FIRE CAPTAIN WRITTEN EXAM STUDY GUIDE

FIRE CAPTAIN ORAL EXAM STUDY GUIDE

XXXXXXXXXXXXXXXXXXXXXXXXXXXXXXXXXXXXXXXXXXXXXXXXXXX

INDEX

INDEX

"A"

ADIABATIC	83
AFFF	126, 129
AFTER THE EXAM	66
AIR	82
AIR: ATMOSPHERIC	92
AMBIENT TEMPERATURE	83
APPARATUS DRIVING	15
ASSESSMENT CENTERS	31, 78
ATMOSPHERIC: AIR	92
ATOMS	80
AUXILIARY COOLER	13

"B"

B.T.U.	88, 89
BACKDRAFT	91, 92
BAFFLE-BOARD	135
BLEVE	113
SIGNS	113
BOIL	89
BOILING POINT	89
BOILOVER	89
BOOSTER PUMP	140
BOYLE'S LAW	86
BREACHING	109
BRISANCE	113
BRITISH THERMAL UNIT	88
BUDGET REVIEW	21
BURNING	
CONDITIONS	90
RATE	89

"C"

CALORIE	83
CANDIDATES	61
CAPACITY: PUMP	14
CAPTAIN	
DUTIES	50
EXAM	21
ORAL	23
PREPARATION	24
QUALIFICATIONS	26
RESPONSIBILITIES	50
WRITTEN EXAM	22
CARRYALL	112
CENTIGRADE	84
CERTIFICATION	10
CHEMICAL: COMPOSITION	126
REACTIONS	113
CORROSIVE	55

"C"
(continued)

CHEMISTRY: FIRE	90
CHURN VALVE	13
CISTERN	116
COMBUSTIBLES	96
COMBUSTION	84, 91, 93
SPONTANEOUS	90
PRODUCTS OF	93
COMMAND POST	38
COMMAND	96
COMPLETE FIREFIGHTER	4, 149
CONDUCTION	80, 90
CONFLAGRATION	103, 104
CONVECTION	81, 90
CORROSIVE CHEMICALS	55

"D"

DEFLAGRATION	113
DETROIT DOOR OPENER	109
DIFFUSION	85
DISTRIBUTION MAINS	115
DISTRIBUTORS	115, 145
DRIVING: APPARATUS	15
DUST	105

"E"

EDUCATION	3, 27
ELEMENTS	80
ENERGY	80
ENGINE COMPANY: FUNCTION	36
ENGINEER	
PRACTICAL	13
QUESTIONS	11
EXAM	11
ENVIRONMENT: WORK	28
ESSAY TEST	54
EXAM	
AFTER	66
CANDIDATES	61
CAPTAIN	21
CAPTAIN WRITTEN	22
ESSAY	54
FILL IN THE BLANKS	57
FORMATS	53
MATCHING	56
MULTIPLE CHOICE	53
ORAL BOARDS	67
PRACTICAL	13
PREPARATION	62
PRINCIPLES	52

"E" (continued)

EXAM:	PROCTORS	52
	SIMULATOR	37, 40
	STRATEGY	52, 62
	TRUE FALSE	54
ENGINEER		11
	ORAL	15
	SIMULATOR	33
	WRITTEN	12
EXPERIENCE		20, 27, 50, 113
EXPLOSION		113
EXPLOSIVE		113
	LIMITS	113, 114
	RANGE	114
EXPOSURE		102, 103
	COVERING	102
	FIRE	102
	PROTECTING	102
EXTINGUISHING AGENT		126
EXTINGUISHING SYSTEMS		124

"F"

FAHRENHEIT		57, 84
FIRE		
	ATTACK	98
	CAUSE	82
	CHEMISTRY	90
	DEVIL	104
	DISCOVERY	96
	ENDURANCE	103
	ENGINEER	24
	EXPOSURE	102
	EXTENSION	103
	EXTINGUISHER	55, 125
	FLOW	117, 118
	FLOW PRESSURE	119
	FUEL	104
	GAS	82
	HOSE	135
	HYDRANT	119
	HYDRANT LOCATION	99
	INDIRECT ATTACK	100, 101
	METAL	126
	MODES	91
	OVERHAUL	110
	PHASE	92
	POINT	87
	PREVENTION	24
	SALVAGE	110, 111
	SCIENCE	80, 81
	SMOLDERING	93
	SPREAD	82
	STAGES	91
	STORM	82
	STRATEGY	95
	STREAM	135, 143
	TACTICS	95
	TETRAHEDRON	81
	TRAVEL	96
	TRIANGLE	81
	VENTILATION	
	WIND	
FIRE APPARATUS SPOTTING		98

"F" (continued)

FIRE CAPTAIN		
	JOB ANNOUNCEMENT	60
FIRE CAPTAIN ORAL EXAM GUIDE		48, 78, 149
FIRE CAPTAIN WRITTEN GUIDE		22, 149
FIRE DEPARTMENT		
	OBJECTIVES	95
	RESPONSE	97
FIRE ENGINEER		
	JOB ANNOUNCEMENT	59
	POSITION	16
	PREPARATION	16
	RESPONSIBILITIES	17, 20
FIRE ENGINEER ORAL GUIDE		17, 149
FIRE ENGINEER WRITTEN GUIDE		12, 149
FIRE EXTINGUISHERS		124
	CARBON DIOXIDE	125
	CHEMICAL	124, 125
	HALOGEN	126
	PORTABLE	125, 126
	RATINGS	124
FIRE HOSE		
	CONSTRUCTION	135
	LAYS	138
	MINIMUM	137
	SPECIFICATIONS	137
FIRE POINT		87
FIRE SERVICE ASSESSMENT		
	CENTERS	78
FIRE STREAM		140, 141
	BROKEN	142
	CAPACITIES	140
	COURSE	142
	CURVATURE	140
	DEFECTIVE	136
	DEFLECTION	144
	DISCHARGE	141
	EFFECTIVE	141
	EFFECTIVENESS	140
	FOG	145
	HEAVY	100
	HORIZONTAL REACH	140
	MASTER	144
	MOBILITY	142
	PENETRATION	144
	SOLID	146
	SPRAY	101, 123, 146
	SPRAY-FOG	144
	STRAIGHT	146
	VERTICAL REACH	140
	MANEUVERABILITY	98
FIREFIGHTER		
	ASSIGNMENTS	9
	ENTRANCE	2
	EXAM CHECK LIST	4
	EXPERIENCE	3, 10
	INVOLVEMENT	25
	JOB PERFORMANCE	8
	KNOWLEDGE	10
	LEARN	2
	PARAMEDIC	23
	QUALIFICATIONS	9
	QUESTIONS	5
	REQUIREMENTS	2
	RESPONSIBILITIES	8
FIREFIGHTER ORAL EXAM GUIDE		5, 149

"F"
(continued)

FIREFIGHTER WRITTEN GUIDE 5, 65, 149
FIREFIGHTERS WRITTEN EXAM 64
FIREFIGHTING 94
    BASIC 96
    FOAM 129
    FORCIBLE ENTRY 109
    INDIRECT ATTACK 100, 101
    MANPOWER 95
    METHODS 97
    OPERATIONS 99
    PUBLIC ASSEMBLY 97
    SEQUENCE 97
    SIZE-UP 64, 94, 95
    STRATEGY 95
    TACTICS 95
    VENTILATION 107
FLAME PROPAGATION 103
FLAME ENVELOPE 81
FLAME 81, 82
FLAMMABLE
    GASES 85
    LIMITS 114
FLASH CARDS 63
FLASH POINT 87
FLASHOVER 82
FLOOR RUNNER 112
FLYERS 58
FOAM 127, 129
    AFFF 129
    ALCOHOL 129
    CONVENTIONAL 129
    ORDINARY 129
    PROPORTIONER 129
    RATIO 129
    SURFACE 129
FORCIBLE ENTRY 109
FORCIBLE ENTRY TOOL
    AIR CHISEL 110
    BATTERING RAM 110
    DETROIT DOOR
        OPENER 109
    FLAT HEAD AXE 110
    PICK-HEAD AXE 109
FRICTION LOSS 146, 148
FUNCTION
    RESCUE COMPANY 36
    SALVAGE COMPANY 36
ENGINE COMPANY 36
    TRUCK COMPANY 36

"G"

GASES 83, 86
    FLAMMABLE 85
GRIDIRON 115

"H"

H2O 55
HALOGEN 127
HALOGENATED AGENT 126

"H"
(continued)

HALON 127
HARD SUCTION 147
HAZARD
    CONFLAGRATION 104
    EXPOSURE 102
HEAT 87, 88
    CAPACITY 87
    COMBUSTION 87
    EXOTHERMIC 88
    RADIATION 12
    SPECIFIC 88
    TRANSFER 89
    VAPORIZATION 87
    LATENT 88
    TRANSMISSION 87
HEATING: SPONTANEOUS 90
HOSE
    BEDS 135
    BRIDGE 137
    COMPARTMENTS 135
    COTTON JACKET 138
    DACRON 139
    DOUBLE JACKET 138
    DOUGHNUT ROLL 137
    HARD SUCTION 147
    LENGTH 143
    LOAD 136
    MULTIPLE JACKET 138
    TESTING 146
    TOOLS 139
HUMIDITY 83
HYDRANT 119
    CAPACITY 119
    VALVE 120
HYDRANTS: DRY 120
HYDROKINETIC 147

"I"

IGNITION
    SPONTANEOUS 90
    TEMPERATURE 86
INCIDENT COMMAND 38
    OPERATIONS 39
INTERVIEW: ORAL 70

"J"

JOB ANNOUNCEMENTS 58
JOB KNOWLEDGE 28
JOB PERFORMANCE 20, 50

"K"

KNOWLEDGE 20, 50

"L"

LADDER COMPANY
- DUTIES 105
  - FUNCTION 105
- LAMINER 119
- LEEWARD 102, 108
- LIEUTENANT PREPARATION 24
- LIGHT WATER 129
- LIQUIDS 86

"M"

- MANAGEMENT REQUIREMENTS 32
- MANAGEMENT SKILLS 32
- MANPOWER: ESTIMATING 40
- MASTER FIRE STREAM 144
- MATCHING QUESTIONS 56
- MATTER 80
- MISCIBILITY 84
- MOLECULE 57, 80, 82
- MONITOR 144
- MOTIVATION 26
- MULTIPLE CHOICE EXAMS 53
- MUSHROOMING 93

"N"

- N.F.P.A. #196 135
- N.F.P.A. #198 135
- N.F.P.A. #1001 9
- NITROGEN 92
- NOZZLE
  - FOG 145
  - PRESSURE 146
  - SIZE 138

"O"

ORAL
- CAPTAIN 23
- EXAM CHECK-LIST 69
- INTERVIEW 70
- NON-STRESSFUL 68
- PREPARATION 74
- STRESSFUL 68
- ORAL BOARDS 68
- ORAL EXAM
  - PRINCIPLES 66
  - AFTER 78
  - PREPARATION 69
- ORGANIZATION 21
- OVERHAUL 110, 111
  - OPERATIONS 110
- OXIDATION 84, 90, 91

"P"

- PH SCALE 84
- PICK-HEAD AXE 109
- PIKE POLE 110
- PLASTICS 105
- PRACTICAL
  - EXAM 13
  - EXAM PREPARATION 15
- PRE-FIRE-PLANNING 96
- PREPARATION: EXAM 62
- PRESSURE
  - EXCESSIVE 140
  - RESIDUAL 119
- PRIMARY FEEDERS 115
- PRINCIPLES
- PRINCIPLES: ORAL EXAM 66
- PROCTOR: EXAM 52
- PUMP CAPACITY 14
- PUMPING
  - AT FIRE 17
  - OPERATIONS 15
  - RELAY 100

"Q"

- QUALIFICATIONS: CAPTAIN 26
- QUESTION: TYPES 58
  - CAPTAIN ORAL 23
  - FILL IN THE BLANKS 57
  - MATCHING 56
  - TRUE FALSE 54
  - WRITTEN 11
  - CAPTAIN 21
  - ENGINEER 11
  - FIREFIGHTER 5

"R"

- RADIATION, ... 81, 90
- REDUCING AGENTS 91
- RELAY PUMPING 100
- REQUIREMENTS: MANAGEMENT 32
- RESCUE
  - OPERATIONS 105
  - PROCEDURES 106
- RESCUE COMPANY
  - FUNCTION 36
- RESERVOIRS. 116
- RESIDUAL PRESSURE 119
- RESPONSIBILITIES
  - CAPTAIN 50
  - FIRE ENGINEER 17, 20
  - FIREFIGHTER 8

"S"

SALVAGE 111
    COVER WORK 112
    COVERS 111
    OPERATIONS 112
SALVAGE COMPANY 111
    FUNCTION 36
SCIENCE
    FIRE 80, 81
SECONDARY FEEDERS 115
SIMULATOR
    EXAM 33, 37, 40
    ROLE PLAYING 35
    TECHNIQUES 34
SIMULATOR EXAM
    COMMANDING INCIDENT 41
    ESTIMATING MANPOWER 40
        EXAMPLE 46
        FORMAT 37
        GRADING SHEET 44
        INCIDENT EXAMPLE 45
        INITIAL REPORT 38, 39, 47
        POST 42
    GRADED 42
    ROUTING 40
    UNUSUAL INCIDENTS 41
SIZE-UP 54, 94
    STAGES 94
SKILLS: MANAGEMENT 31, 32, 82
SPARK 82
SPECIFIC GRAVITY 84
SPECIFIC HEAT 88
SPONTANEOUS 90
    COMBUSTION 90
    IGNITION 90
SPRINKLER
    COVERAGE 131
    RATINGS 131
SPRINKLER SYSTEMS 130, 133
    AUTOMATIC 130, 131
    COMBINED 132
    CONNECTION 131
    DELUGE 132, 133
    DRY PIPE 132
    PRE-ACTION 131, 132
    PRE-ACTION DELUGE 132
    SUPERVISORY 131
    TYPES 130
    WET PIPE 132
STANDPIPE
    SIAMESE 134
STANDPIPE SYSTEMS 134, 135
    CLASS I 134
    CLASS II 134
    CLASS III 134
    TYPES 134
STEAM 86
STRAINER 11
STRATEGY: EXAM 52
    DIVISIONS 30
SUBLIME 86
SUBSTANCE 86
SUPPLEMENTARY
    WATER SUPPLIES 120
SURFACE TENSION 128

"T"

TASK: IMPROPER 30
TEMPERATURE 84
THEORY 95
THERMOSYPHON 121
TRAINING 20, 26, 50
TRUCK COMPANY
    FUNCTIONS 36
    OPERATIONS 105
TRUE FALSE QUESTIONS 54
TURBULENCE 121

"V"

VALVE
    CHURN 13
    HYDRANT 120
        INDICATING 120
        OS&Y 120, 133
        PIV 120, 133
VAPOR 85
    DENSITY 85
    PRESSURE 85
VENTILATION 106, 107
    ARTIFICIAL 107
    CROSS 107, 108
    FORCED AIR 108
    PRINCIPLE 106
    REASONS FOR 107
VISCOUS WATER 128

"W"

WATER 55, 84, 98, 99, 121, 122
    CONSUMPTION 116, 117
    COOLING AGENT 122
    DAMAGE 103
    FOG 123
    FREEZING POINT 121
    LIGHT 128
    MAIN 114, 121
    PIPES 121
    REQUIREMENTS 116
    SUPPLY 114, 116
    SUPPLY POINT 117
    SYSTEM 114, 117, 118
    TABLE 116
    WET 127
WATER SYSTEM
    ADEQUACY 116
    CAPACITY 118
    DIRECT PUMPING 115
    PUBLIC 116
    RELIABILITY. 116
WATER FORM 121
WATERWAY 121
WEATHER 104
WET-WATER 128
WETTING AGENTS 127, 128
WINDWARD 108
WORK ENVIRONMENT 28
WRITTEN EXAM 12, 22, 54

FIRE SERVICE STUDY GUIDES

FOR

FIRE CAPTAIN

FIRE LIEUTENANT

FIRE   ENGINEER

FIRE PUMP OPERATOR

FIRE APPARATUS DRIVER

ENTRANCE LEVEL FIREFIGHTER

ADVANCEMENT IN THE FIRE SERVICE

FIRE SERVICE EXAM CLASSIFICATIONS:

BACKGROUND   INVESTIGATIONS

PHYSICAL AGILITY EVENTS

MEDICAL EXAMINATIONS

ASSESSMENT CENTERS

ORAL INTERVIEWS

SIMULATIONS

PRACTICAL

WRITTEN

ABSOLUTELY GUARANTEED!

The information contained in these **STUDY GUIDES** will give **YOU** the **POSITIVE EDGE** in the **FIRE SERVICE EXAM** process. These **STUDY GUIDES** are **ORGANIZED** so that **YOU** can cover a maximum amount of **VARIED** information without having to spend **VALUABLE** study time searching for the required information.

YOU WILL COVER MORE INFORMATION IN LESS TIME! IF NOT FULLY SATISFIED RETURN BOOK/BOOKS FOR FULL REFUND !

BOOKS AVAILABLE FROM INFORMATION GUIDES ARE LISTED ON THE FOLLOWING PAGES, ALONG WITH AN ORDER FORM:

BOOK #1

"FIRE ENGINEER WRITTEN EXAM STUDY GUIDE" 2nd. ed.
$15.95 per copy

180 page book with 10 chapters containing over **2500** selections of information that **ALL FIRE FIGHTERS** should know. Each piece of information is presented in a **NEW** and **UNIQUE METHOD** that will permit **EFFICIENT - ORGANIZED** studying. A must book that will assist **ANY FIREFIGHTER** in obtaining a high score on the **WRITTEN** portion of the **FIRE ENGINEERS** promotional exam.
Perfect bound soft cover: 5 1/2" X 8 1/2"
ISBN 0-938329-52-9 LCCN 86-81239

BOOK #2

"FIRE ENGINEER ORAL EXAM STUDY GUIDE" 2nd ed.
$15.95 per copy

192 page book with seven chapters containing over **400 PRACTICAL - ORAL INTERVIEW** and **SITUATION** type questions. Each question is followed by a **SUGGESTED** response. The **MOST COMPLETE** book of it's kind. An **ABSOLUTE MUST** as a tool to prepare **ANY FIRE FIGHTER** for the **PRACTICAL - ORAL** portion of the **FIRE ENGINEER - APPARATUS DRIVER - PUMP OPERATOR** promotional exam.
Perfect bound soft cover: 5 1/2" X 8 1/2"
ISBN 0-938329-53-7 LCCN 88-80888

BOOK #3

"FIRE CAPTAIN WRITTEN EXAM STUDY GUIDE"
$18.95 per copy

288 page book with ten chapters containing over **3000** selections of information **ALL FIRE FIGHTERS** should know. Each bit of information is presented in a **NEW/UNIQUE METHOD** that will permit **EFFICIENT** and **ORGANIZED** studying. This is a **MUST BOOK** that will assist **ANY FIRE FIGHTER** in obtaining a high score on the **WRITTEN** portion of the **FIRE CAPTAIN-LIEUTENANT** promotional exam.
Perfect bound soft cover: 5 1/2" X 8 1/2"
ISBN 0-938329-54-5 LCCN 88-80890

BOOK #4

"FIRE CAPTAIN ORAL EXAM STUDY GUIDE"
$18.95 per copy

228 page book with ten chapters containing over **500 ORAL INTERVIEW** and **SITUATION** type questions. Each question is followed by a **SUGGESTED** response. The most complete book of its kind. An **ABSOLUTE MUST**, as a tool that will prepare **ANY FIRE FIGHTER** for the **ORAL** portion of the **FIRE CAPTAIN - LIEUTENANT** promotional exam.
Perfect bound soft cover: 5 1/2" X 8 1/2"
ISBN 0-938329-55-3 LCCN 88-80889

BOOK #5

"THE COMPLETE FIREFIGHTER CANDIDATE"
$12.95 per copy

150 page book with eight chapters containing the **MOST COMPLETE** inventory of information available. This book will guide prospective Firefighters through all the essential steps that need to be taken in order to become the **COMPLETE FIREFIGHTER CANDIDATE**.
Perfect bound soft cover   5 1/2" X 8 1/2"
ISBN 0-938329-58-8   LCCN 89-81738

BOOK #6

"FIREFIGHTER WRITTEN EXAM STUDY GUIDE"
$19.95 per copy

336 page book with 12 chapters containing over 3000 selections of information - questions - answers that **ALL PROSPECTIVE FIREFIGHTERS** should know. Each selection of information is presented in a **UNIQUE METHOD** that will permit **EFFICIENT** and **ORGANIZED** studying. A must book that will assist Firefighter candidates in obtaining a high score on the **WRITTEN** portion of the **FIREFIGHTERS ENTRANCE EXAM**.
Perfect bound soft cover: 5 1/2" X 8 1/2"
ISBN 0-938329-59-6   LCCN 89-81736

BOOK #7

"FIREFIGHTER ORAL EXAM STUDY GUIDE"
$15.95 per copy

226 page book with eight chapters containing over **400 ORAL INTERVIEW** and **SITUATION** type questions. Each question is followed by a suggested response. The **MOST COMPLETE** book available for preparing Firefighter candidates for the ORAL portion of the **FIREFIGHTERS ENTRANCE EXAM**.
Perfect bound soft cover: 5 1/2" X 8 1/2"
ISBN 0-938328-61-8   LCCN 89-81737

BOOK #8

A SYSTEM FOR: " ADVANCEMENT IN THE FIRE SERVICE"
$9.95 per copy

170 page book with six chapters containing all the essential steps and procedures that Firefighters should follow in order to promote within the **FIRE SERVICE**. This book is presented in a **THOROUGH** and **ORGANIZED** manner so as to allow Firefighters to see the overall picture for **ADVANCEMENT IN THE FIRE SERVICE**.
Perfect bound soft cover: 5 1/2" X 8 1/2"
ISBN 0-938329-56-1   LCCN 88-083385

BOOK #1
FIRE ENGINEER WRITTEN EXAM

Fire apparatus, fire prevention, fire pumps, fire streams, fire behavior, fire hydraulics, fire prevention, tools and equipment, water supply, hazardous materials, fire extinguishing systems.

BOOK #2
FIRE ENGINEER ORAL EXAM

Oral interview preparation, job knowledge, personal information general knowledge, and actual situation questions and responses.

BOOK #3
FIRE CAPTAIN WRITTEN EXAM

Fire administration, training, fire fighting fire prevention, fire behavior, fire apparatus and equipment, fire extinguishing systems, fire streams, water supply, and hazardous materials.

BOOK #4
FIRE CAPTAIN ORAL EXAM

Oral interview preparation, job knowledge, personal information, general knowledge, situation questions, incident simulator preparation, assessment centers.

BOOK #5
COMPLETE FIREFIGHTER CANDIDATE

Introduction to the Fire Service, Fire Department familiarization, exam check list, practice exams, locating exams, job announcements, job applications, resumes, exam process, exam divisions, what to prepare for, written/oral/agility/medical exam information and much more.

BOOK #6
FIREFIGHTER WRITTEN EXAM

Types of exams and questions, general aptitude and judgement, reading, vocabulary, spelling, grammar, science, mechanical comprehension, physics, chemistry, pattern analysis, math, progressions, Fire Service information, first aid, pre-test study books, and much more!

BOOK #7
FIREFIGHTER ORAL EXAM

Oral interview preparation, job knowledge, general knowledge, personal knowledge, actual situation questions and responses.

BOOK #8

ADVANCEMENT IN THE FIRE SERVICE
Prior to obtaining a position as a Firefighter, after obtaining a position as a Firefighter, after promoting to the position of Engineer, after promoting to the position of Captain.

ORDER FORM

NAME _____

FIRE DEPARTMENT _____

STREET ADDRESS _____

CITY _____

STATE AND ZIP _____

PHONE NUMBER (_____)_____

NUMBER OF BOOKS:

#1___ $15.95 = $_____.____   #5___ $12.95 = $_____.____

#2___ $15.95 = $_____.____   #6___ $19.95 = $_____.____

#3___ $18.95 = $_____.____   #7___ $15.95 = $_____.____

#4___ $18.95 = $_____.____   #8___ $ 9.95 = $_____.____

CALIFORNIA RESIDENCE ADD 6.5% SALES TAX:
(.065 times the total price of books ordered)

SALES TAX = $_____

SHIPPING & HANDLING CHARGES:

BOOK RATE MAIL:
$1.00 for the 1st copy; $0.50 each additional copy.

RUSH FIRST CLASS MAIL:
$3.00 for the 1st copy; $1.00 each additional copy.

SHIPPING AND HANDLING = $_____

MAKE CHECKS or M.O. OUT TO : "INFORMATION GUIDES"

TOTAL AMOUNT ENCLOSED = $_____

CREDIT CARD ORDERS : _____ VISA _____ MASTER CARD

CREDIT CARD NUMBER : __ __ __ __ __ __ __ __ __ __ __ __

EXPIRATION DATE : _____

NAME ON CARD : _____

SIGNATURE : _____

**RUSH ORDERS PHONE: 1-800 "FIRE BKS" = 1-800-347-3257
ALSO: (213) 379-1094**

BOOK REVIEWS OF "STUDY GUIDES"

FIREFIGHTER NEWS:

"Every Firefighter that has been or is currently involved in the Fire Service exam process will attest to the need for these books."

REKINDLE MAGAZINE:

"Information is presented in a new and unique method for efficient and organized studying."

FIREHOUSE MAGAZINE:

"Well thought-out study books for the person aspiring to promote."

FIRE CHIEF MAGAZINE:

"Information that Firefighters - Candidates should know in preparing for Fire Service exams."

FIRE TRAINING OFFICER:

"I used the books as a reference for a portion of our Fire Department exams, very useful."

GRANVILLE, TEXAS - FIREFIGHTER:

"There is a need for more books of this type!"

FIRE ENGINEERS JOURNAL MAGAZINE:

"Stimulates the reader to think about relevant facts presented in a two or three line summary, which can be retained in memory!"

THE INSTITUTION OF FIRE ENGINEERS:

"These books are a benefit to all Firefighters - Candidates taking Fire Service examinations."

FIREFIGHTER ESCONDIDO, CALIFORNIA:

"Excellent books, excellent format, excellent content, very useful."

YUBA CITY - FIRE EQUIPMENT OPERATOR:

"Very useful and informative. Need more books like these."

COLLEGE FIRE SCIENCE INSTRUCTOR:

"Books are very useful as reference guides for all of my classes!"